Brendan

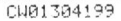

Brendan Long grew up in Western Canada and has had a lifelong interest in nature and wildlife conservation. He has been providing advice in respect of energy investments for over 20 years, covering everything from traditional energy to hydrogen fuel cells. He has studied at Bishop's University, the London School of Economics, the Institut d'Etudes Politiques de Paris (Sciences Po) and Texas A&M University. He is a Certified Financial Analyst (CFA) Charterholder.

Copyright © Brendan Long 2021

All rights reserved. No part of this book may be reproduced or used in any manner without written permission of the copyright owner except for the use of quotations in a book review.

ISBN 979-8-7145-4364-7

Contents

Introduction: Prometheus 1

Part 1: The Natural History of Fire 2

  1. The First Flames 3

    From Fire: Carbon Dioxide 3

    The First Fire on Earth 5

    From Fire: Homo Sapiens 7

  2. Wildfires 11

    Wildfires in North America 12

    Peatfires and Peat 16

    Wildfires in Australia 17

    Wildfires in Rainforests? 20

  3. Wildfires Lit by Humans 22

    Wildfires in the Great Plains 23

    Wildfires in Africa 25

    Wildfires in the Amazon 30

    Shrinking Wildfires: Shrinking Wilderness 33

  4. Farmland: Wildfire Boundary 35

    Farmland: the Basis of Civilization 35

    Farmland: Taken from Wilderness 40

    Farmland: to End Hunger 43

    Farmland: a Cornerstone 46

Part 2: From Fire: Human Development 48

  1. Fire Starting: Our First Tech 49

  2. From Fire: Materials 51

    From Fire: Bricks, Glass and Cement 51

From Fire: the Bronze Age 57

   From Fire: the Iron Age 62

3. From Fire: Energy... 69

   Fire from Wood, Energy for the Billions....................... 70

   Fire from Coal, the Embattled King 72

   Fire from Oil, to Win 75

   Fire from Natural Gas, Blue Flames 81

   Fire from Bio-fuel, Farmed Fuel................................... 84

   Alternatives to Fire 85

      Hydro Power.. 86

      Nuclear Power 86

      Wind Power 87

      Solar Power.. 88

      Electrical Batteries... 90

      Hydrogen 92

      Costs of Fire vs. Alternatives to Fire 94

   Global Energy Supply by Source.................................. 96

Part 3: Extinguishing the Human Use of Fire........................ 98

   1. From Fire: Pollution 99

   2. From Fire: Global Warming 102

   3. From Fire: Carbon Dioxide Fertilization 108

   4. From Fire: Politics 118

Fire: Summary and Conclusion 126

Introduction: Prometheus

According to Greek mythology, Prometheus stole fire from the gods and gave it to humans. He observed that each of the other animals had an advantage, but humans had none. Where were our claws to protect us, our wings to fly or our furs to keep us warm? Prometheus took it upon himself to give us an advantage: fire.

With fire we were on the up. From our lowly and perilous origins, the advantage of fire elevated humankind to greatness.

From our earliest beginnings, fire has provided us with warmth, light and cooked food. Fire allowed us to make the materials upon which we built civilization, namely, bricks, glass, bronze and steel. Fire, today, provides nine-tenths of our global needs in energy.[1]

Humankind is choosing to extinguish the gift of Prometheus because the burning of fire emits carbon dioxide.

Fire was written to provide thought provoking perspectives by tracing the history of humanity and of our Earth through the storyline of fire.

Fire challenges conventional thinking and brings the lessons of history to the heart of the issues that matter today: addressing global warming, ending global hunger and protecting wildlife.

This is the story of fire.

---

[1] **World Energy Balances 2020 (Data for 2018)** – International Energy Agency – July 2020

Part 1: The Natural History of Fire

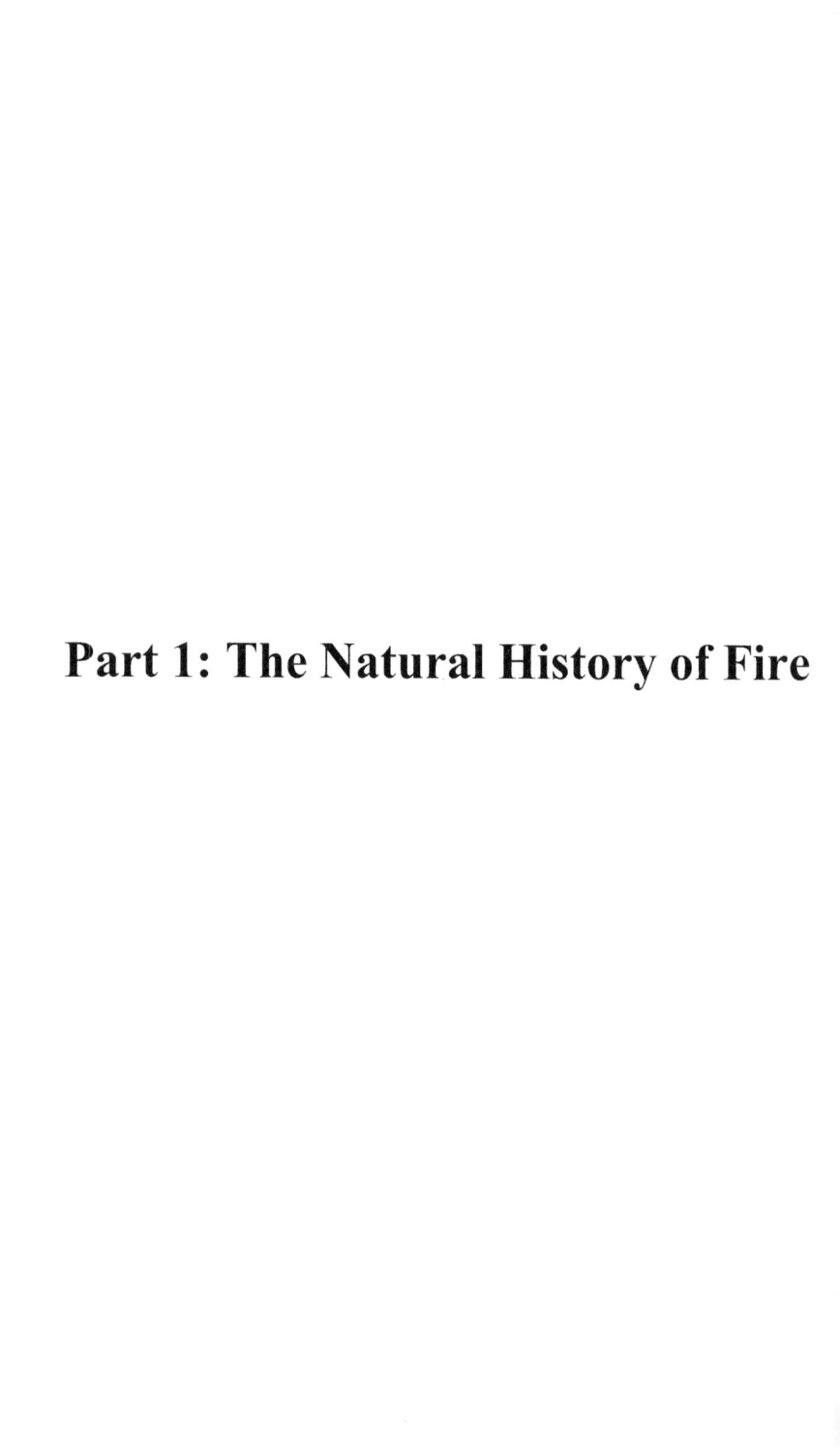

1. The First Flames

"Three, two, one, zero, all engines running, lift-off, we have a lift-off."[2]

Fire may have propelled us to great heights and even to the moon, but what is it?

Fire propelling the launch of Apollo 11 — *Figure 1*

Photo Credit: NASA

From Fire: Carbon Dioxide

Fire is a chemical reaction that combines fuel with oxygen to produce heat.

The release of energy in the form of heat is the quintessential characteristic of fire.

Fire can take many forms.

Within an internal combustion engine fire occurs in a highly controlled environment to produce heat. That heat causes air to

---

[2] **Apollo, Humankind's first steps on the lunar surface, Sounds from Apollo 11** – NASA – https://www.nasa.gov – Accessed: March 2021

expand with an explosive force. The force of that expanding air powers all internal combustion engines.

Fire can also occur in uncontrolled environments. Wildfires that burn forests and grasslands are prominent examples.

Although fires can take an infinite number of forms, they all burn fuel and the energy in all fuel is derived, at its origins, from sunlight. Revealingly, according to mythology, Prometheus stole fire from Helios, the god of the sun.

Fire reverses the processes that create organic matter – plants and vegetation. The burning of kerosene to launch a rocket,[3] the burning of a forest fire and the burning of coal to create electricity – all fires – burn organic matter.

Plants use the energy from the sun to combine water, carbon dioxide and other molecules to create organic matter. Much of the energy required to create organic matter is retained within it.

Fire releases the energy retained within organic matter in the form of heat and turns organic matter back into water, carbon dioxide and other molecules.

Carbon is the universal building block of life. Due to the organic origins of fuel, it contains carbon. As a result, fire produces carbon dioxide. Carbon dioxide is causing global warming. This will be discussed further in the section "From Fire: Global Warming". Due to global warming, humanity is seeking to stop the human use of fire.

There is one important fuel that burns differently from all the rest: hydrogen. Hydrogen fuel is the product of synthetic, inorganic processes designed by humans. Hydrogen does not contain carbon and burning it does not emit carbon dioxide. Burning hydrogen produces water and nothing else – other than heat. Setting aside theoretical exceptions, the energy stored in hydrogen fuel created by humans is derived, at its origins, from sunlight.

---

[3] **The F-1 Engine Powered Apollo Into History, Blazes Path for Space Launch System Advanced Propulsion** – NASA – https://www.nasa.gov – Accessed: March 2021

Flames occur when the heat released by fire causes molecules to become so hot that they radiate light.[4]

The First Fire on Earth

No one can say with certainty when the first fire erupted on our Earth. However, we do know that there are three requirements for wildfires to burn: heat, oxygen and fuel.[5] By ascertaining when all three of these requirements first coexisted, we can deduce when the first fires burned on Earth.

Lightning is the primary source of heat that ignites naturally occurring wildfires. Lightning is created by turbulent air movements within clouds that mix liquid and frozen water in such a way that it creates an electric charge. If the charge becomes great enough, it overcomes the resistance of the atmosphere and an electrical current between the cloud and the Earth is formed – lightning.[6]

The Earth is the only planet in our universe that is known to produce lightning in this way.[7] On Earth, the presence of clouds is the essential precondition for the formation of lightning. Our Earth has had creeks, rivers, clouds, rain and lightning for the last 3.8 billion years.[8] Let us now turn our attention to oxygen, the second precondition for the occurrence of fire.

As it turns out, for the first two billion years of the Earth's existence, oxygen molecules were not present in any meaningful capacity in our atmosphere. Oxygen molecules (O_2) are chemically unstable and

---

[4] **Soot: Giver and Taker of Light** – Christopher Shaddix and Timothy Williams – American Scientist – May/June 2007
[5] **Elements of Fire** – US Department of Agriculture Forest Service – https://smokeybear.com – Accessed: March 2021
[6] **What is lightning? Severe Weather 101 FAQ** – The National Severe Storms Laboratory – https://www.nssl.noaa.gov – Accessed: July 2020
[7] **Lightning on Venus inferred from whistler-mode waves in the ionosphere** – C. T. Russell, T. L. Zhang, M. Delva, W. Magnes, R. J. Strangeway and H. Y. Wei – Nature – 29 November 2007; and
NASA Scientist Confirms Light Show on Venus – Christopher Russell, D.C. Agle, Dwayne Brown, Monica Talevi – NASA Mission News – 28 November 2007 – www.nasa.gov – Accessed: July 2020
[8] **The Water Cycle** – Steve Graham, Claire Parkinson and Mous Chahine – NASA Earth Observatory – 1 October 2010 – https://Earthobservatory.nasa.gov – Accessed: July 2020

tend to break apart to form other molecules, which is why, initially, oxygen was scarce in our atmosphere. It was only around 2.3 billion years ago that oxygen rather suddenly – perhaps over a period of 10 million years – burst onto the scene in a big way. This development is known as the Great Oxidation Event. Organisms are thought to have lived on Earth for at least one billion years before this event,[9] but the ascendance of one organism changed everything: cyanobacteria. Cyanobacteria are single celled organisms that live in water and appear as blue-green algae. Through the process of photosynthesis, cyanobacteria use energy from sunlight to transform carbon dioxide and water into complex organic matter. In addition to creating organic matter, photosynthesis produces molecular oxygen (O_2) as a by-product.

Photosynthesis and cyanobacteria predate the Great Oxidation Event, but during that event cyanobacteria took off and began producing oxygen molecules at a much faster rate than oxygen molecules could be depleted by natural oxidation processes (inclusive of nocturnal plant processes that absorb oxygen from the atmosphere). Therefore, it was during this time that oxygen began to accumulate in our atmosphere. For the last 2.3 billion years, photosynthesis has produced oxygen in quantities such that it has existed in significant concentrations in our atmosphere.[10]

Having established when the first and second requirements for fire became present on Earth, let us turn our attention to fuel, the third requirement for fire.

Before the Silurian Period (444-419 million years ago), plants had grown only in water, in which all things are relatively weightless. Starting in the Silurian Period, plants evolved to grow outside of water by supporting their own weight against gravity. This continued into the following Devonian Period (419-359 million years ago), during which the first trees and forests emerged on our planet – fuel.

Plants evolved to grow upwards and to support their weight against gravity by taking advantage of the attributes of carbon atoms. Carbon-based molecules form the rigid cell structures that, combined

---

[9] **The Secret of How Life on Earth Began** – Michael Marshall – BBC – 31 October 2016
[10] **The History of Air** – Riley Black – The Smithsonian – 18 April 2020

together, allow plants to grow upwards against the constant pull of gravity. Fire transforms the carbon in organic matter into carbon dioxide, which escapes as a gas into the atmosphere.

We have established that all three preconditions for wildfires, namely, heat, oxygen and fuel, have been present on Earth from the Silurian Period to the present time. We can therefore conclude that fires have occurred on Earth from the Silurian Period to the present.[11]

Having determined when the first fires erupted on Earth, let us look at when humans first used fire.

From Fire: Homo Sapiens

In contrast to widely held beliefs, humans did not invent the use of fire. Fire was used by now-extinct species long before our species existed.[12]

Humans have been using fire to cook food since we have existed as a species, starting some 300,000 years ago. Most importantly, advances in science indicate that the human species is genetically dependent on fire for its survival: Human morphology and in particular our energy-consuming brains and small digestive systems reflect that we require fire and the cooked food it provides.[13] Cooked food provides significantly more energy, net of digestion, than

---

[11] **Charcoal in the Silurian as evidence for the earliest wildfire** – Ian Glasspool, Dianne Edwards and Lindsey Axe – Geology – May 2004

[12] **Microstratigraphic evidence of in situ fire in the Acheulean strata of Wonderwerk Cave, Northern Cape province, South Africa** – Francesco Berna *et al.* – Proceedings of the Natural Academy of Sciences of the United States of America (PNAS) – 15 May 2012

[13] **Catching Fire: How Cooking Made Us Human** – Richard Wrangham – Profile Books – September 2009;
Control of Fire in the Palaeolithic, Evaluating the Cooking Hypothesis – Richard Wrangham – Current Anthropology – 16 August 2017 and
Food for Thought: Was Cooking a Pivotal Step in Human Evolution? The dietary practice coincided with increases in brain size, evidence suggests – Alexandra Rosati – Scientific America – 26 February 2018

uncooked food.[14] Cooked food and human morphology allow us to allocate energy away from digesting towards thinking.[15]

Advances in science over the last two decades suggest that without fire humans would not exist.[16]

Humans were not the first species to be dependent on fire. The extinct, Homo erectus, whose name means "upright human", takes that accolade. Like humans, they had large brains relative to their body size, small teeth, reduced jaws, reduced chins and slim stomach cavities, as evidenced by their narrow rib cage.[17]

The oldest definitive archaeological evidence of the use of fire for cooking is dated to 1.0 million years ago,[18] or roughly 700,000 years before our species, Homo sapiens, emerged.

Our tastes reflect that we have acquired genetic traits that have encouraged us to eat cooked food due to the benefits it provides. Until the Agricultural Revolution some 12,500 years ago, cooking consisted of grilling. Grilling involves building a fire for the purposes of creating embers over which food can be cooked.[19] Embers emit a dry heat and very little smoke or soot, which disflavors food. Grilling with embers produces direct thermal radiation that heats the exterior of food to temperatures over 150° Celsius (300° Fahrenheit). Under these conditions, the flavor of food changes through processes

---

[14] **Cooking shapes the structure and function of the gut microbiome** – Rachel N. Carmody *et al.* – Nature Microbiology – 30 September 2019

[15] **Catching Fire: How Cooking Made Us Human** – Richard Wrangham – Profile Books – September 2009

[16] **Catching Fire: How Cooking Made Us Human** – Richard Wrangham – Profile Books – September 2009;
Control of Fire in the Palaeolithic, Evaluating the Cooking Hypothesis – Richard Wrangham – Current Anthropology – 16 August 2017 and
Food for Thought: Was Cooking a Pivotal Step in Human Evolution? The dietary practice coincided with increases in brain size, evidence suggests – Alexandra Rosati – Scientific America – 26 February 2018

[17] **Control of Fire in the Palaeolithic, Evaluating the Cooking Hypothesis** – Richard Wrangham – Current Anthropology – 16 August 2017 and
Homo erectus, our ancient ancestor – Lisa Hendry – Natural History Museum – https://www.nhm.ac.uk – Accessed: July 2020

[18] **Microstratigraphic evidence of in situ fire in the Acheulean strata of Wonderwerk Cave, Northern Cape province, South Africa** – Francesco Berna *et al.* – Proceedings of the Natural Academy of Sciences of the United States of America (PNAS) – 15 May 2012

[19] **The Barbecue Bible** – Stephen Raichlen – Workman Publishing Company – 1998

known as Maillard Reactions, after the Frenchman who in 1912 first explained their chemistry. Maillard Reactions turn food into golden and brown colors. The compounds produced by these reactions give the flavor to foods such as caramel, roasted coffee, the cheese on baked pizza and beer, which is the product of roasted barley.[20] Many of the most appetizing foods for humans reflect that we are genetically adapted to prefer food that has been cooked by fire, or more specifically, food that has been grilled over embers.

Humans are defined as a species by our use of fire. We will see in the section "From Fire: Materials" that fire has been instrumental in producing the basic materials on which our civilizations have been built. We will see in the section "From Fire: Energy" that fire has given motion to our creations.

Fire, however, is not the only attribute that distinguishes humans from all other species. Humans are distinct relative to all other species due to our use of fire *and* our use of language.

Human morphology has not changed since we emerged as a species some 300,000 years ago. We have been big-brained from the beginning. However, it is thought that we acquired language genetically only 80,000 years ago, in "barely a blink of an eye in evolutionary time, presumably involving some slight rewiring of the brain."[21] The dating of the human acquisition of language is based on the sudden proliferation of art, symbols and jewelry from that time.[22]

Although the definition of what constitutes language remains controversial, the most accepted current thinking is that language is a genetically acquired ability to connect discrete linguistic units together through a codified process so that they have meaning.[23]

---

[20] **An Introduction to the Maillard Reaction: The Science of Browning, Aroma, and Flavor** – Eric Schulze – seriouseats.com – 13 April 2017 – Accessed: March 2020
[21] **Why Only Us Language and Evolution** – Robert C. Berwick and Noam Chomsky – The MIT Press – 2016
[22] **Why Only Us Language and Evolution** – Robert C. Berwick and Noam Chomsky – The MIT Press – 2016
[23] **The Language Myth** – Vyvyan Evans – Cambridge University Press – October 2014;
The Unfolding of Language – Guy Deutscher – Holt Paperbacks – January 2005;
The truth about language: what it is and where it came from – Michael Corballis – The University of Chicago Press – 2017; and

Although many animals communicate amongst each other, none, other than humans, is capable of building complex thoughts with linguistic units. The longest sentence ever communicated via sign language by Nim, an orangutan trained from birth to communicate like a human, was "Give orange me give eat orange me eat orange give me eat orange give me you."[24] That sentence is considered to be the equivalent of the random hitting of buttons on a dispensing machine in the hope that a refreshment might fall out. Linguists are in agreement that this does not represent language.

The Theory of Evolution is a masterwork that explains how an almost infinite number of living organisms fit perfectly into the puzzle of our natural world.[25] However, the defining characteristics of humans fit uneasily into that puzzle. We are adapted genetically not just to our natural world, but to our mastery of fire – an invention. Additionally, language was acquired too quickly to be consistent with that theory[26] and it vastly supersedes what can be explained by utilitarian, evolutionary reasoning.[27]

Humans won the evolutionary jackpot twice: firstly, by becoming genetically adapted not just to our natural world, but to fire – an invention – and, secondly, by the genetic acquisition of language. No other living species has either of these genetic attributes, and humans have both.

Our acquisition of fire and language explains why humans are so different from the other animals with whom we cohabitate our planet and why the course of human development is unlike anything else that has ever occurred on our Earth.

Having gained an understanding of fire and of our own natural history, it is time to focus on the physical manifestation of fire in our natural world, wildfires.

Why Only Us Language and Evolution – Robert C. Berwick and Noam Chomsky – The MIT Press – 2016
[24] **Researcher Challenges Conclusion That Apes Can Learn Language** – Dava Sobel – The New York Times – 21 October 1979
[25] **On the Origin of Species by Means of Natural Selection** – Charles Darwin – 1859
[26] **Why Only Us Language and Evolution** – Robert C. Berwick and Noam Chomsky – The MIT Press – 2016
[27] **Language and Wallace's Problem** – Stephen Levinson – Science – 27 June 2017

2. Wildfires

4.4% of the land area that supports plant growth on Earth burns annually based on the data from NASA's Terra satellite (4.2 million square kilometers per year; 1.6 million square miles per year).[28]

There are three types of wildfires:[29]

i) Crown fires burn trees from the bottom to the top. They are the most intense and dangerous type of wildfire.

ii) Surface fires burn grasses, shrubs, leaves, pine needles, dead branches and fallen trees.

iii) Underground wildfires burn peat. We will assess these fires in the section "Peatfires and Peat". Through natural geological processes buried peat is transformed into coal. Therefore, peat will also be discussed in the sections "Fire from Coal, the Embattled King" and "From Fire: Global Warming".

Wildfires are also of interest because the United Nations Intergovernmental Panel on Climate Change determined "with very high confidence" that global warming is causing extreme wildfires that are causing harm to humans and ecosystems.[30]

The carbon dioxide emitted by the human use of fire, namely, fire from burning coal, oil and natural gas, is causing global warming. Therefore, wildfires are also relevant to understanding the justifications being put forward to stop the human use of fire. This will be further discussed in the section "From Fire: Politics".

---

[28] **The Global Fire Atlas of individual fire size, duration, speed and direction** – Niels Andela, Douglas C. Morton, Louis Giglio, Ronan Paugam, Yang Chen, Stijn Hantson, Guido R. van der Werf and James T. Randerson – Earth System Science Data – 24 April 2019

[29] **Fire Behaviour** – Natural Resources Canada – www.nrcan.gc.ca – Accessed: February 2020

[30] **Climate Change 2014, Synthesis Report** – Editors: Rajendra K. Pachauri, Leo Meyer and Core Writing Team – Fifth Assessment Report of the Intergovernmental Panel on Climate Change – 2015

Wildfires in North America

In contrast to widely held beliefs that nature is balanced, stable and repeating, wildfires are random and chaotic. They also dominate many ecosystems.

Every year there are thousands of wildfires in North America. However, most wildfires are of little significance. It is the exceptions, the rare but large wildfires, that burn substantially all the surface area burned by wildfires.[31]

To fully appreciate the natural character of wildfires requires us to go back in time to a period of transition during which humans kept wildfire records while having a limited suppressive influence on them. The Miramichi Fire of 1825 burned in both the Canadian province of New Brunswick and the bordering American state of Maine.[32] That fire burned for only two days before being extinguished by heavy rain, yet it burned an area equal to that of the state of New Jersey, USA – equivalent to about half the surface area of Switzerland.[33]

For over a decade preceding the Miramichi Fire, the climate of the area was cold and wet, which suppressed wildfires and allowed fuel loads to build up. However, the year of the fire itself was particularly hot and dry. Additionally, in the year of the fire, a pest had killed many trees in the area, which dried them out and made them highly flammable.[34]

The winds created by the fire were reported to be of a hurricane force, reaching speeds of 110 kilometers per hour (70 miles per hour). The

---

[31] **Forest Fires** – Natural Resources Canada – www.nrcan.gc.ca – Accessed: February 2020
[32] **The international nature of the Miramichi Fire** – Alan MacEachern – The Forestry Chronicle – May/June 2014
[33] **State Area Measurements** – United States Census – https://www.census.gov – Accessed: February 2021; and
Surface Area Switzerland – Word Bank – https://data.worldbank.org – Accessed: February 2021
[34] **The Great Miramichi Fire** – Brian Bouchard – Maine Genealogical Society – 22 October 2009

flames from the fire were reported to move forwards at speeds exceeding 100 kilometers per hour (60 miles per hour).[35]

The Miramichi Fire of 1825 remains the most devastating fire in Canadian history with fatality estimates ranging from 200 to 500 people.[36] For reference, the Peshtigo Fire of October 1871 killed at least 1,200 people and remains the most devastating fire in the history of the United States.[37]

Although wildfires are actually beneficial to and required by many ecosystems,[38] due to their destructive and dangerous nature, wildfires have been actively suppressed by humans.

The Wildfire Prevention Campaign of the United States is the longest-running public service advertising campaign in American history. However, it is representative of a global wildfire prevention campaign that has been running for much longer. Specifically, wildfires have been suppressed by sedentary people for the last 12,500 years or since the beginning of the Agricultural Revolution. This will be discussed in the section "Farmland: Wildfire Boundary".

The Wildfire Prevention Campaign of the United States gained prominence due to its use of the Disney character Bambi, a mule deer fawn, as seen in Figure 2.[39]

---

[35] **The Great Miramichi Fire** – Brian Bouchard – Maine Genealogical Society – 22 October 2009
[36] **Forest Fires** – The Canadian Encyclopaedia – https://thecanadianencyclopedia.ca – Accessed: June 2020
[37] **The Great Midwest Wildfires of 1871** – Tom Hultquist – US National Weather Service – https://www.weather.gov/grb/peshtigofire – Accessed: June 2020
[38] **Wildland Fire in Ecosystems, Effects of Fire on Fauna** – Jane Kapler Smith – U.S. Department of Agriculture, Forest Service, Rocky Mountain Research Station – 2000; and
Wildfires as an ecosystem service – Juli G Pausas and Jon E Keeley – Frontiers in Ecology and the Environment – 6 May 2019
[39] **History of Smokey Bear** – Legacy Document, South Dakota Department of Agriculture – Accessed via https://sdda.sd.gov – Accessed: June 2020

Wildfires presented as a threat to Bambi *Figure 2*

Image Credit: US Department of Agriculture, Forest Division

The success of Bambi inspired the Wildfire Prevention Campaign to develop its own character in 1944, Smokey Bear, as seen in Figure 3.[40]

We must prevent wildfires *Figure 3*

Image Credit: US Department of Agriculture, Forest Division (Archives U. of Illinois)

In 1999, the Colorado Legislature had a report prepared by the Colorado Division of Wildlife to understand the reasons for which mule deer populations were falling. Surprisingly, wildfire

---

[40] **The story of Smokey Bear** – US Forest Service, Office of Communication – 4 August 2014

suppression was determined to be a direct cause of the decline in mule deer populations.[41] How is that possible?

Mule deer have light digestive systems that are adapted to eating the young lush plants that grow in the clearings made in forests by wildfires.[42]

Paradoxically, the suppression of wildfires, intended to protect Bambi, deprived Bambi's real-life kin of their food and contributed to a reduction in mule deer populations.

As an important digression, it is critical to appreciate that well-intentioned actions that do not consider the complete set of consequences of those actions can have an effect that is opposite to the one intended. Governments, businesses, consumers and citizens around the world are taking well-intentioned measures to stop the human use of fire, namely, fire from burning coal, oil and natural gas. However, those actions generally ignore the most important impact that will result from the replacement of fire with alternatives to fire, namely, how that change will affect farming. This will be developed in the sections "Farmland: Wildfire Boundary" and "From Fire: Human Development".

In the decades that followed the creation of Smokey Bear, North Americans continued their longstanding pursuit of suppressing wildfires. However, from the 1970s, there has been a growing awareness that focusing exclusively on wildfire suppression is ineffective and also detrimental to ecosystems.[43]

Ecological learnings and the teachings of the First Peoples of North America advocated that wildfires were actually required in many ecosystems. Today, the first heading on the Smokey Bear website is entitled "The Benefits of Fire". According to Smokey Bear and the

---

[41] **Declining Mule Deer Populations in Colorado: Reasons and Responses, A Report to the Colorado Legislature** – Prepared by: R. Bruce Gill with Contributions from: T. D. I. Beck, R. H. Kahn, C. J. Bishop M. W. Miller, D. J. Freddy, T. M. Pojar, N. T. Hobbs G. C. White. – November 1999

[42] **Mule deer habitat** – Utah Division of Wildlife Resources – https://wildlife.utah.gov – Accessed: March 2020

[43] **Fire Management, An Attitude Shift** – Natural Resources Canada – https://www.nrcan.gc.ca – Accessed: March 2021

US Forest Service, fires "trigger a rebirth of forests, helping to maintain native plant species."[44]

Peatfires and Peat

Peatfires and peat have become subjects of international interest due to the capacity of peatlands to store carbon.

The relevance of peatlands will resurface in the section "Fire from Coal, the Embattled King" because peat turns into coal through natural geological processes. Understanding peatlands is also instructive for the section "From Fire: Global Warming" because burning coal is causing global warming. Effectively, the story of global warming begins here with peatfires and peat.

So, what is peat?

Peat is essentially the accumulation of dead organic material that does not decay. Peaty soil is acidic and has a low oxygen content, which prevents decomposition. Undecomposed organic matter can constitute between 80%-100% of peat. The settling of organic materials on peatlands can create peaty soils that have thicknesses of 10 meters (33 feet) or more.

If peaty soils become dry, they can burn. Decomposition and fires both result in carbon combining with oxygen to form carbon dioxide, which escapes as a gas into the atmosphere.

Peatfires can be some of the hardest wildfires to extinguish. They can smolder through the winter even when buried under snow.[45]

In total, 3.3% of the Earth's land area consists of peatlands, of which 91% is located in the northern latitudes. These northern peatlands started forming about 15,000 years ago as the Earth slowly began to exit the last ice age. On average since the end of the last ice age, dead organic matter containing 18.1 grams of carbon has settled on each square meter of peatland every year (0.58 ounces per square yard). That may not sound like a lot, but consider that the Earth has 4.4

---

[44] **Benefits of Fire** – US Forest Service – https://www.smokeybear.com – US Forest Service – Accessed: June 2020

[45] **Peatland fires and carbon emissions** – Natural Resources Canada – www.nrcan.gc.ca – Accessed: January 2021

million square kilometers (1.7 million square miles) of peatland. Peatlands have grown to hold 612 billion metric tons of carbon.[46]

Currently, the carbon being captured per year by peatlands equates to the equivalent of 370 million metric tons of carbon dioxide.[47] That figure equates to an offset of 1.1% of the carbon dioxide emitted by global human energy use in 2019.[48] Peatlands are the largest terrestrial store of carbon.

When a wildfire burns a forest or grassland, the fire releases carbon dioxide. Carbon dioxide is subsequently recaptured when the forest or grassland grows back, forming a natural carbon cycle. This is not the case for peatfires.

Peatlands and peatfires are both completely at odds with our conception of nature being balanced and cyclical. The rich organic matter consumed by peat is simply lost in the Earth – gone. Perhaps buried peat might by some fluke resurface in the form of coal, the product of buried peat, in some forthcoming geological period; that would relate to geological circumstances, not to an organic cycle. Likewise, the carbon dioxide emitted by peatfires is simply lost to the atmosphere. There is not a counterbalancing natural mechanism to recapture that carbon dioxide.

We believe that nature is balanced and cyclical; however, in respect of the carbon that is captured by peatlands and the carbon that is released by peatfires nature is neither balanced nor cyclical.

Wildfires in Australia

Although wildfires play an integral role in many of the Earth's ecosystems, Australia's wildfire-dependent ecosystems are in a category of their own. The eucalyptus trees that dominate Australian ecosystems are not only adapted to defend themselves against

---

[46] **Global peatland dynamics since the Last Glacial Maximum** – Zicheng Yu, Julie Loisel, Daniel P. Brosseau, David W. Beilman, Stephanie J. Hunt – Geophysical Research Letters – 9 July 2010

[47] **Peatlands and climate change** – International Union for the Conservation of Nature – November 2017

[48] **Global CO2 emissions in 2019** – International Energy Agency – 11 February 2020

wildfires, they proactively promote them by regularly shedding highly flammable bark and leaves.[49]

Unburned, this fuel builds up on the ground until, eventually, it ignites when there is a wildfire. The more fuel that has built up, the more intense the wildfire will be.

The map in Figure 4 highlights regions of the world where wildfire burn areas fell over the 1998-2015 period. The entire north and north-eastern eucalyptus-dominated forests and grasslands of Australia are colored on that map. How do you expect a prolonged period of wildfire suppression in a eucalyptus-tree dominated ecosystem is likely to have ended?

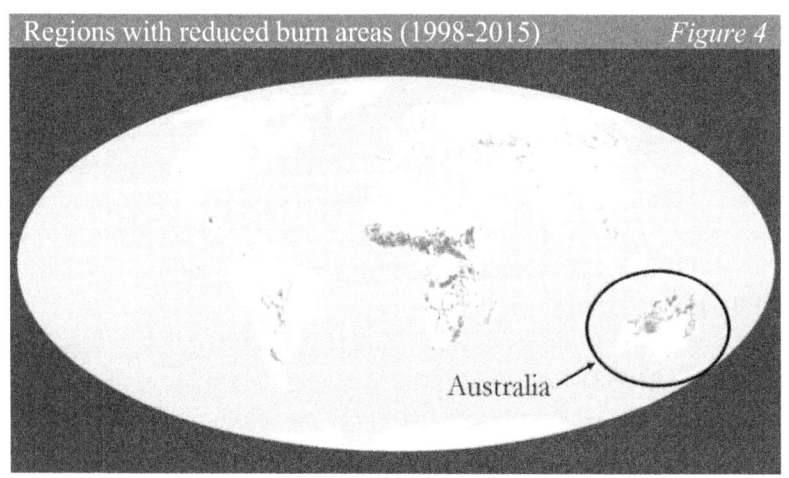

Figure 4: Regions with reduced burn areas (1998-2015)

Image Credit: Adapted from NASA[50]

The 2019-2020 summer in Australia was known as the black summer due to the extraordinary plumes of smoke that wafted over the country. Tragically, at least 33 people were killed by the blazes including 4 firefighters.[51]

---

[49] **Eucalyptus Globulus** – US Department of Agriculture Forest Service – https://www.fs.fed.us – Accessed: March 2021
[50] **NASA Detects Drop in Global Fires** – Image Credit: Joshua Stevens, NASA's Earth Observatory – https://www.nasa.gov – Accessed: February 2021
[51] **Australia fires: A visual guide to the bushfire crisis** – BBC News – 31 January 2020

As an important digression relating to eucalyptus trees, it is important to understand the extent to which eucalyptus trees are transforming landscapes outside of Australia.

In the mid-19th century, Portugal desperately needed more timber and was concerned by the erosion caused by deforestation. What better tree to import than the eucalyptus tree of Australia? They are aesthetic, fast-growing, strong and have medicinal oils in their leaves. Today, non-native eucalyptus trees cover a surface area of some 9,000 square kilometers (3,500 square miles) in Portugal.[52]

The rapid settling of California, USA, resulted in the equally rapid deforestation of its oaks and redwoods. As a result, in 1868, California passed the Tree Culture Act to incentivize tree planting. At the time, Australia's eucalyptus trees were getting raving reviews. Moreover, the fast-growing eucalyptus trees matched the needs of a fast-growing state. Today, non-native eucalyptus trees dominate much of California, as seen in Figure 5.

Spread of non-native eucalyptus trees in California — *Figure 5*

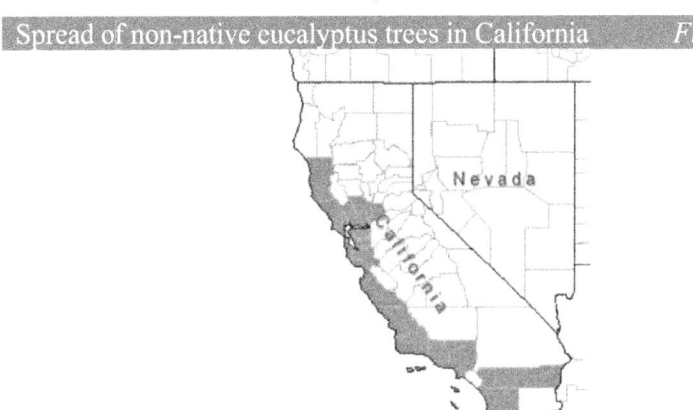

Image Credit: US Forest Service[53]

Wildfire-related news has been dominated by fires in Australia (1999-2020), which caused at least 33 fatalities; Portugal (2017),

---

[52] **Historical Development of the Portuguese Forest: The Introduction of Invasive Species** – Leonel J. R. Nunes, Catarina I. R. Meireles, Carlos J. Pinto Gomes and Nuno M. C. Almeida Ribeiro – Forests – 4 November 2019; and
Portugal's 'killer forest' – Paul Ames – Politico – 19 June 2017
[53] **Eucalyptus Globulus, Fire Effects Information System** – US Forestry Service – https://www.fs.fed.us – Accessed: October 2020

which caused at least 100 fatalities; and California, USA, (2018) which caused at least 87 fatalities.[54]

Australia, Portugal and California all have landscapes that are dominated by eucalyptus trees. Moreover, due to the proximity of these eucalyptus-tree dominated areas to population centers, wildfire suppression tactics have been applied in each of those regions. Combined, this is a formula for deadly, high-intensity wildfires.

Landscape transformations caused by humans and wildfire suppression strategies are a root cause of extreme and dangerous wildfires.

In the section "From Fire: Carbon Dioxide Fertilization" we will discuss that science has only recently become aware of the extent to which the changing vegetative landscapes of the Earth – forests and grasslands – affect the climate. The transformation of landscapes due to the spread of non-native vegetation initiated by human interference is an example of both the complexity of the Earth's climate systems and the extent to which human influence is affecting our Earth.

Wildfires in Rainforests?

Wildfires have profound impacts on most vegetative ecosystems, but what role do wildfires play in tropical rainforests?

There is an abundance of evidence suggesting that wildfires are unnatural in tropical rainforests. Tropical rainforests are simply too humid and wet.[55]

---

[54] **Portugal fires: Interior Minister resigns as death toll passes 100, PM under fire to stand down** – Reuters/AP – 19 October 2017;
California wildfires death toll climbs to 87, almost 500 still unaccounted for – Morgan Winsor – ABC News 24 November 2018; and
Australia fires: A visual guide to the bushfire crisis – BBC – 31 January 2020
[55] **Holocene fire and occupation in Amazonia: records from two lake districts** – Mark B Bush, Miles R Silman, Mauro B de Toledo, Claudia Listopad, William D Gosling, Christopher Williams, Paulo E de Oliveira and Carolyn Krisel – The Royal Society Publishing – 9 January 2007; and
A 6,000 year history of Amazonian maize cultivation – Mark B. Bush, Dolores, R. Piperno and Paul A. Colinvaux – Nature – 27 July 1989

Scientists only recently discovered the underlying reason for which tropical rainforests are so humid. This research started in the Amazon rainforest and is taking on a global dimension involving all forests.[56]

It has been recently determined that the trees of the Amazon river basin transpire more water into the atmosphere than the Amazon River drains into the Atlantic Ocean.[57] The trees of the Amazon create their own rain. The entire Amazon rainforest emits 1.37 meters (4.5 feet) of water per year into the atmosphere – rain in reverse.[58]

We will see in the section "From Fire: Global Warming" that water vapor has been recognized as being *by far* the Earth's most important greenhouse gas since the discovery of greenhouse gases in 1859. The significance of water vapor from trees as a cause of global warming will be discussed in the section "From Fire: Carbon Dioxide Fertilization" after we assess the unexpected and extraordinary scale of the *growth* in the Earth's forested surface area, net of losses, based on the data from NASA's Terra satellite. The key point to retain is that scientists are just beginning to understand the scale of the water vapor that is emitted by trees and vegetation – science is changing and fast.

Returning to the subject of wildfires, human activity is so overwhelmingly dominant on wildfires it has been difficult in this section to assess their natural character. In the next section we will focus directly on wildfires lit by humans.

---

[56] **Forests, atmospheric water and an uncertain future: the new biology of the global water cycle** – Douglas Sheil – Forest Ecosystems – 20 March 2018

[57] **The largest river on Earth is invisible and airborne** – Dan Kedmey – Science – 24 November 2015; and
Study shows the Amazon makes its own rainy season – Carol Rasmussen – NASA's Jet Propulsion Laboratory – 17 July 2017

[58] **The land–atmosphere water flux in the tropics** – Fisher JB, Malhi Y, Bonal D, Da Rocha HR, De Araújo AC, Gamo M, Goulden ML, Rano TH, Huete AR, Kondo H, Kumagai T, Loescher HW, Miller S, Nobre AD, Nouvellon Y, Oberbauer SF, Panuthai S, Roupsard O, Saleska S, Tanaka K – Global Change Biology – (15) 2009

3. Wildfires Lit by Humans

Contrary to prevailing expectations, the annual amount of surface area burned globally by wildfires *decreased* by 24.3% over the 18-year period ending in 2015 based on data from NASA's Terra satellite.[59]

Both the direction and the scale of the change were massive surprises. What is going on?

Nomadic people light wildfires because doing so creates good grazing land for the animals that they hunt. In contrast, sedentary or farm-based societies suppress wildfires because they are a threat to the property of those societies.

The satellite data reflects that the last of the Earth's nomadic societies are becoming agricultural.

Citing from NASA, "Using satellites to detect fires and burn scars from space, researchers have found that an ongoing transition from nomadic cultures to settled lifestyles and intensifying agriculture has led to a steep drop in the use of fire for land clearing and an overall drop in natural and human-caused fires worldwide."[60]

NASA's Niels Andela indicated, "As soon as people invest in houses, crops and livestock, they don't want these fires close by anymore."[61]

---

[59] **A human-driven decline in global burned area** – N. Andela, D. C. Morton, L. Giglio, Y. Chen, G. R. van der Werf, P. S. Kasibhatla, R. S. DeFries, G. J. Collatz, S. Hantson, S. Kloster, D. Bachelet, M. Forrest, G. Lasslop, F. Li, S. Mangeon, J. R. Melton, C. Yue, J. T. Randerson – Science – 30 June 2017

[60] **Researchers Detect a Global Drop in Fires** – Kate Ramsayer – NASA Earth Observatory – https://earthobservatory.nasa.gov – Accessed: June 2020

[61] **Researchers Detect a Global Drop in Fires** – Kate Ramsayer, citing Niels Andela – NASA Earth Observatory – https://earthobservatory.nasa.gov – Accessed: June 2020

Wildfires in the Great Plains

The Great Plains cover about one quarter of the North American continent and extend from the Arctic tundra to Texas, USA.[62] I grew up in the Great Plains of Canada and remember being driven 800 kilometers (500 miles) across them every summer to visit my grandparents' farm near the city of Melfort, Saskatchewan. After crossing an endless sea of grass, arriving at my grandparents' farm, the first thing I would see was, not grass, but a giant elm tree about 12 meters high gently shading their farmhouse. That tree has always been puzzling. If trees can grow in the Great Plains, why are they grasslands and not forests?

Wildfires are the reason the Great Plains and fertile grasslands are barren of trees. If you suppress wildfires, trees grow in wild grasslands so long as they get enough water. Advances in forestry research are suggesting that the humidity created by forests creates rain that allows forests to expand into dry areas.[63] Wildfires push back on that expansion.

Wildfires in grasslands are caused by lightning and by humans. Nomadic hunter-gatherers lit wildfires in the Great Plains to create grazing land for the animals that they hunted.[64]

The benefits of wildfires for nomadic hunters are abundantly clear as soon as we step into wild grasslands, which in many cases requires us to step back in time:

In 1804, Meriwether Lewis set off from St. Louis, Missouri, to lead a successful US Federal Government sponsored expedition to the Pacific Ocean and back. According to his journal entry made on the 16th of September 1804 while crossing the Great Plains (and including the antiquated spellings):

---

[62] **Geography: The Great Plains** – US National Parks Service – https://www.nps.gov – Accessed: February 2021
[63] **How Forests Attract Rain: An Examination of a New Hypothesis** – Douglas Sheil and Daniel Murdiyarso – BioScience – April 2009
[64] **References on the American Indian Use of Fire in Ecosystems** – Gerald W. Williams – United States Department of Agriculture, Forest Service – 18 May 2005 – https://www.nrcs.usda.gov – Accessed: February 2021

> "[T]hese extensive planes had been lately birnt and the grass had sprung up and was about three inches high. Vast herds of buffaloe deer elk and antelopes were seen feeding in every direction as far as the eye of the observer could reach."[65]

As an important digression, as a singular example of the changes that are taking place in respect of climate science, in 2018, a team of leading scientists suggested that human lit wildfires in the Great Plains would have affected the climate by changing that region's forested surface area.[66] The acceptance by science that changes to the vegetative landscapes of the Earth affect the climate represents a transformational, and recent, development. This will be further discussed in the section "From Fire: Carbon Dioxide Fertilization".

The transformation of the Great Plains from wild and fiery grasslands into farmland coincided with extraordinary losses of large mammals across the plains. The bison population in North America before the settlement of that continent by Europeans is estimated to have been between 30 and 60 million.[67] The bison population fell precipitously over the course of the 19th century. Realizing the scale of the losses, in 1887, the American Museum of Natural History, based in New York, sent an expedition to the state of Montana to obtain one of the last remaining specimens – none was found.[68] Today, thanks to the efforts of grassroots conservationists, the North American bison population has grown from near-extinction to about 350,000.[69]

Wildfires, apex predators and large disruptive mammals have a lot in common: They are dangerous, difficult to contain, wide-ranging and can behave in ways that are difficult to predict. Additionally, when humans permanently settle in an area, they all tend to be eradicated.

---

[65] **Journals of the Lewis and Clark Expedition** – University of Nebraska – https://lewisandclarkjournals.unl.edu – Accessed: February 2021
[66] **Indigenous impacts on North American Great Plains fire regimes of the past millennium** – Christopher I. Roos, María Nieves Zedeño, Kacy L. Hollenback and Mary M. H. Erlick – Proceedings of the National Academy of Sciences of the United States of America – 7 August 2018
[67] **Timeline of the American Bison** – United Stated Fish and Wildlife Service – Accessed: June 2020
[68] **Timeline of the American Bison** – United Stated Fish and Wildlife Service – Accessed: June 2020
[69] **Meet the American Bison** – The Nature Conservancy – https://www.nature.org – Accessed January 2021

Neither wildfires, nor apex predators, nor large disruptive mammals have a presence to speak of in the lives of most settled people. For reference, settled people can be considered to include all people whose food is derived from agriculture as opposed to hunting and gathering.

Prior to the settlement of the United States by Europeans it is estimated that more than half of the country contained large areas in which fire occurred at least once every twelve years.[70] It was not only the Great Plains that burned: The tree rings of a 2,210-year-old tree from Sequoia National Park, California, indicated that it experienced, on average, a fire every 18 years over its lifetime.[71]

Using ten years of data up to and including 2019, the implied average frequency of wildfires across the Lower-48 states of the United States is one fire every 352 years.[72]

If the world is a fiery place today, it is a lot less so. This is because nomadic hunter-gatherer societies have been displaced by settled agricultural societies that suppress wildfires.

In contradiction to widely held beliefs, from the perspective of maintaining the natural ecosystems of North America, the greatest threat relating to wildfires is their absence.

Wildfires in Africa

The First Peoples of North America have been shaping ecosystems with wildfires since inhabiting that continent starting about 15,000 years ago, a couple millennia before the end of the last ice age.[73]

---

[70] **Presettlement Fire Frequency Regimes of the United States** – Cecil C. Frost – Pages 70-81 in Teresa L. Pruden and Leonard A. Brennan (eds.). Fire in Ecosystem Management: Shifting the Paradigm from Suppression to Prescription. Tall Timbers Fire Ecology Conference Proceedings, No. 20. Tall Timbers Research Station, Tallahassee, FL. – 1998

[71] **Multi-Millennial Fire History of the Giant Forest, Sequoia National Park, California, USA** – Thomas W. Swetnam, Christopher H. Baisan, Anthony C. Caprio, Peter M. Brown, Ramzi Touchan, R. Scott Anderson & Douglas J. Hallett – Fire Ecology – 1 December 2009

[72] **National Interagency Fire Center** – https://www.nifc.gov – Accessed: July 2020

[73] **The First Americans** – Scientific American – 1 November 2012; and
The Fertile Shore – Fen Montaigne – The Smithsonian Magazine – January / February 2020

Likewise, the First Peoples of Australia have been shaping ecosystems with wildfires since inhabiting that continent starting about 65,000 years ago.[74] Let us go further back in time, much further back. It is time to take a look at Africa.

In total, Africa's grasslands cover around half of that continent, albeit estimates vary depending on how grasslands are defined.

Grasses first grew on Earth 35 million years ago. They started dominating African landscapes 10 million years ago,[75] which is consistent with the timing of the expansion of grasslands elsewhere on Earth.[76]

As an important digression, the authors of the study that dated the expansion of African grasslands believe that i) grasses are advantaged relative to trees when atmospheric carbon dioxide levels are low and ii) a collapse in atmospheric levels of carbon dioxide 10 million years ago triggered the expansion of grasslands at the expense of forests.[77] If that was the case, human-caused increases in atmospheric levels of carbon dioxide would have the opposite effect and favor the expansion of forests. This will be revisited in the section "From Fire: Carbon Dioxide Fertilization". Of note, that research was published in 2019 – the pace at which science is changing is astounding.

---

[74] **Human occupation of northern Australia by 65,000 years ago** – Chris Clarkson, Zenobia Jacobs, Ben Marwick, Richard Fullagar, Lynley Wallis, Mike Smith, Richard G. Roberts, Elspeth Hayes, Kelsey Lowe, Xavier Carah, S. Anna Florin, Jessica McNeil, Delyth Cox, Lee J. Arnold, Quan Hua, Jillian Huntley, Helen E. A. Brand, Tiina Manne, Andrew Fairbairn, James Shulmeister, Lindsey Lyle, Makiah Salinas, Mara Page, Kate Connell, Gayoung Park, Kasih Norman, Tessa Murphy & Colin Pardoe – Nature – 20 July 2017

[75] **Synchronous rise of African C4 ecosystems 10 million years ago in the absence of aridification** – Pratigya J. Polissar, Cassaundra Rose, Kevin T. Uno, Samuel R. Phelps & Peter deMenocal – Nature Geoscience volume – 22 July 2019

[76] **Grassland fire ecology has roots in the late Miocene** – Allison T. Karp, Anna K. Behrensmeyer, Katherine H. Freeman – Proceedings of the National Academy of Sciences – November 2018

[77] **Synchronous rise of African C4 ecosystems 10 million years ago in the absence of aridification** – Pratigya J. Polissar, Cassaundra Rose, Kevin T. Uno, Samuel R. Phelps & Peter deMenocal – Nature Geoscience volume – 22 July 2019

The first hominids able to walk on two feet emerged in Africa six million years ago (Sahelanthropus).[78] Walking is more useful in ecosystems dominated by grasslands.

Once grasslands are established, the regular occurrence of grass fires, caused by both lightning and hominids, makes it difficult for trees to survive in the area. In particular, saplings are easily destroyed by wildfires. Figure 6 provides an image of trees and grasses co-existing in the iconic Serengeti nature reserve of Tanzania. There, as elsewhere, wildfires tip the balance of nature in favor of grasslands.

Wildfires promote grasslands at the expense of trees   *Figure 6*

Photo Credit: Marcel Kocačič on Unsplash (Location: Serengeti, Tanzania)

Today, almost all wildfires in Africa are known to be started by humans. This is known for many reasons, the most obvious of which is that almost all African wildfires occur in periods of dryness when there is very little lightning. But when did humans first start to light wildfires?

By applying their understanding of the differences between dry-season fires lit by hominids and wet-season fires lit by lightning,

---

[78] **Walking Upright** – Smithsonian, National Museum of Natural History – https://humanorigins.si.edu – Accessed: November 2020

scientists determined that hominid-lit wildfires have been the norm in Africa for at least the last 200-300 thousand years.[79]

Clearly, humans have been shaping landscapes and ecosystems with wildfires for long enough to make it very tricky to disentangle what is natural from what is not.

Science is now accepting that landscape changes and changes to the Earth's vegetation change the climate; by implication, human-lit wildfires would have had a profound effect on the climate of our ancestors. This will be developed further in the section "From Fire: Carbon Dioxide Fertilization" when we assess the amount of water emitted by trees.

To assess current changes in wildfire regimes and their implications for wildlife habitat we will focus on a small fraction of Africa's grasslands; specifically, we will focus on the iconic Serengeti-Masai Mara nature reserve.

The Serengeti National Park is in Tanzania and to the north of that park the adjoining Masai Mara National Reserve is in Kenya.

Earth scientists undertook a study based on satellite data over the period from 2001 to 2014 to assess changes to wildfire dynamics in the Serengeti-Masai Mara. The results are indicative of the speed with which profound changes to our Earth are occurring: Over that relatively short 13-year period, the number of wildfires per year fell by 40% and the amount of burned surface area per year fell by 39%.[80]

The Earth scientists assessed whether other changes detectable by satellite could explain the reduced wildfire activity: They saw that the number of illegal livestock enclosures being used by cattle farmers within the protected areas of the Serengeti-Masai Mara almost doubled over the period of study.

---

[79] **Evolution of human-driven fire regimes in Africa** – Sally Archibald, A. Carla Staver and Simon A. Levin – Proceedings of the National Academy of Sciences of the United States of America – 17 January 2012

[80] **Anthropogenic modifications to fire regimes in the wider Serengeti-Mara ecosystem** – James R. Probert, Catherine L. Parr, Ricardo M. Holdo T. Michael Anderson Sally Archibald, Colin J. Courtney Mustaphi, Andrew P. Dobson, Jason E. Donaldson, Grant C. Hopcraft, Gareth P. Hempson, Thomas A. Morrison, Colin M. Beale – Global Change Biology – 8 July 2019

Based on the satellite data, wildfires were 7.5 times more frequent in the core of the protected area compared to its perimeter. The data indicates that the nomadic custom of lighting wildfires to maintain the grassland ecosystem of the Serengeti-Masai Mara is being displaced by encroaching farmland.

Let us assess the implications of these changes for the area's wildlife.

The Serengeti-Masai Mara provides habitat for many of Africa's best-known animals, including elephants, giraffes, hippos, buffalos, zebras, wildebeests, black rhinos, lions, cheetahs and hyenas.

Historically, elephants did not inhabit the Serengeti-Masai Mara but have sought refuge there due to habitat loss elsewhere.[81] The African elephant population is currently around 415,000. That represents a decline by a factor of more than 20 times since 1930 when 10 million elephants inhabited Africa.[82]

The World Wildlife Fund has indicated that lions inhabit only 8% of the land they once occupied in Africa.[83] As a result, the 3,000 lions that now inhabit the Serengeti-Masai Mara represent 15% of the African lion population.[84]

Satellite analysis is showing the degree to which the ancient practice of prescribed burning is in decline in the Serengeti-Masai Mara. This change is associated with the expansion of farming at the expense of wildlife habitat. Sadly, the area's wild grasslands are increasingly overgrazed by cattle, occupied by crops or simply fenced off.[85]

---

[81] **History Serengeti National Park** – The Serengeti National Park – https://www.serengetiparktanzania.com – Accessed February 2021
[82] **The Status of African elephants** – World Wildlife Fund Magazine – Winter 2018 – https://www.worldwildlife.org – Accessed: January 2021
[83] **The Magnificent Lion: The Symbol of Africa** – World Wildlife Fund – https://www.wwf.org.uk/ – Accessed: November 2020
[84] **Africa: Tanzania Has Largest Number of Lions in Africa, New Report Says** – Edward Qorro – https://allafrica.com/ – Accessed: November 2020 (citing the research of Dr Dennis Ikanda, Principal Research Officer in charge of Carnivore Ecology and Conservation at the Tanzania Wildlife Research Institute)
[85] **Threats to the Serengeti** – Andrew M. Sugden – Science – 29 March 2019; and **Cross-boundary human impacts compromise the Serengeti-Mara ecosystem** – Michiel P. Veldhuis, Mark E. Ritchie, Joseph O. Ogutu, Thomas A. Morrison, Colin M. Beale, Anna B. Estes, William Mwakilema, Gordon O. Ojwang, Catherine L. Parr, James Probert, Patrick W. Wargute, J. Grant C. Hopcraft, Han Olff – Science – 29 March 2019

Changing wildfire regimes are at the heart of this issue, more surprisingly, so too is the human use of fire. Why?

From the Bronze Age, the human use of fire has been the driving force that has increased the efficiency of farming. The more efficiently we use farmland, the more land is available for wildlife. This will be developed in the sections "Farmland: Wildfire Boundary" and "From Fire: Human Development".

The Serengeti-Masai Mara and the Amazon rainforest are both located in the Tropics. For reference, the Tropics represent the area between the tropic of Cancer (latitude of 23.3° North) and the tropic of Capricorn (latitude of 23.3° South). The relevance of this will be developed later when we discuss that both extreme poverty and the destruction of wildlife habitat are concentrated in the Tropics. Without leaving the Tropics, let us now turn our attention to the Amazon rainforest.

Wildfires in the Amazon

By some estimates, humans have inhabited the Amazon for the last 13,000 years.[86] Charcoal from wildfires starts to feature in the sedimentary record of the Amazon exactly with the emergence of South America's staple crops, corn and manioc, some 6,000 years ago – well before the arrival of Europeans.[87] The First Peoples of the Amazon used fire extensively for agricultural purposes.[88]

---

[86] **The Supposedly Pristine, Untouched Amazon Rainforest Was Actually Shaped By Humans** – Ben Panko – Smithsonianmag.com – 3 March 2017
[87] **Holocene fire and occupation in Amazonia: records from two lake districts** – Mark B Bush, Miles R Silman, Mauro B de Toledo, Claudia Listopad, William D Gosling, Christopher Williams, Paulo E de Oliveira and Carolyn Krisel – The Royal Society Publishing – 9 Jan 2007;
A 6,000 year history of Amazonian maize cultivation – Mark B. Bush, Dolores, R. Piperno and Paul A. Colinvaux – Nature – 27 July 1989; and
Ancient farmers burned the Amazon, but today's fires are very different – Kate Evans – National Geographic – 5 September 2019
[88] **Ancient farmers burned the Amazon, but today's fires are very different** – Kate Evans – National Geographic – 5 September 2019

Wildfires lit by humans in the Amazon rainforest would both clear land of trees and increase the fertility of the soil for agriculture.[89] This practice is common amongst the First Peoples who inhabit tropical rainforests around the world and it involves the felling and drying of trees before setting them alight.[90]

Traditionally, the First Peoples of the Amazon rainforest i) practiced the small-scale clearing of patches of treed land, ii) used agricultural land on a rotational basis, which allowed the forest to regrow periodically, iii) left some trees growing in cleared-out areas and iv) planted mixed crops.[91] The current industrial-scale felling and burning of Amazonian trees and industrial agricultural practices are inconsistent with the ancient practices of the Amazon.

Deforestation trends in the Amazon rainforest of Brazil are provided in Figure 7. According to NASA, more than one-sixth, or more than about 17%, of the Amazon rainforest has been deforested due to the expansion of farmland.[92] The deforestation of the Amazon is causing large-scale loss of wildlife habitat and, correspondingly, of wildlife.

---

[89] **The changing chagras: traditional ecological knowledge transformations in the Colombian Amazon** – Valentina Fonseca-Cepeda, C. Julián Idrobo and Sebastián Restrepo – Ecology and Society – Vol 24, Number 1, 2019;
For many tribes in the Amazon, fire is part of their livelihood and culture – Jayalaxshmi Mistry – The Independent – 4 September 2019; and
The role of Amazonian anthropogenic soils in shifting cultivation: learning from farmers' rationales – André B. Junqueira, Conny J. M. Almekinders, Tjeerd-Jan Stomph, Charles R. Clement, Paul C. Struik – Ecology and Society – Volume 21, Number 1, 2016
[90] **These Farmers Slash and Burn Forests, But in a Good Way** – Gleb Raygorodetsky – National Geographic – 8 March 2016; and
Geography of the Third World – C.G Clarke, Dr J P Dickenson, J.P Dickenson, W.T.S Gould, S Mather, Sandra Mather, Prof R Mansell Prothero, R.M Prothero, D.J Siddle, C.T Smith, Mr C T Smith, E. Thomas-Hope – Routledge – 1996
[91] **The legacy of 4,500 years of polyculture agroforestry in the eastern Amazon** – S. Yoshi Maezumi, Daiana Alves, Mark Robinson, Jonas Gregorio de Souza, Carolina Levis, Robert L. Barnett, Edemar Almeida de Oliveira, Dunia Urrego, Denise Schaan & José Iriarte – Nature Plants – 23 July 2018; and
Traditional Land Use and Shifting Cultivation, The Amazon Basin Forest – Global Forest Atlas – Yale School of Forestry
[92] **Making Sense of Amazon Deforestation Patterns** – Adam Voiland – NASA Earth Observatory – https://earthobservatory.nasa.gov – Accessed: October 2020

Amazon deforestation rate (km² per year) *Figure 7*

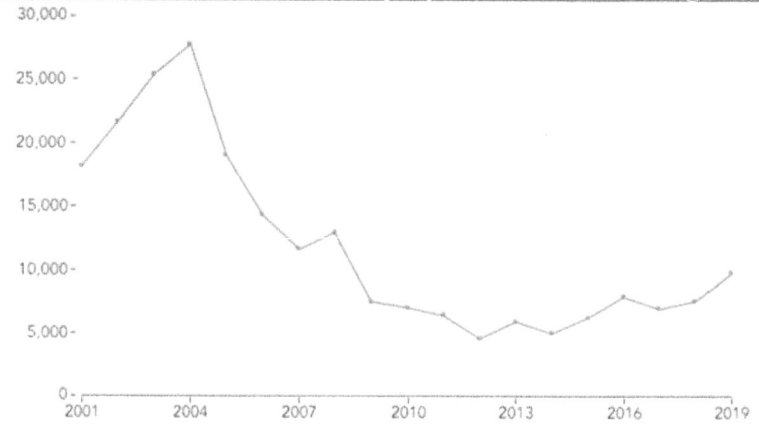

Image Credit: NASA, PRODES[93]

The degree to which the last vast wild places on Earth are being tamed by agricultural pressures is perhaps most dramatically evidenced by the pressures on the uncontacted tribes of the Amazon. As their name implies, these tribes have had no contact with the agricultural societies that are increasingly dominating the region. For reference, there are about 30 uncontacted tribes currently living in the Amazon.[94]

Diseases such as the common cold, flu, measles and chicken pox can kill more than half of these tribes when they are first contacted by outsiders because they have no immunity.[95]

In the 1980s and 1990s, farmers settled remote areas in the state of Rondônia in western Brazil. During this period, many uncontacted tribespeople are thought to have been expelled from their land or killed. In Rondônia there is a plot of forested land of about 40 square kilometers (15 square miles) that is under the protection of the Brazilian government. It is surrounded by farmland. Within it a

---

[93] **Making Sense of Amazon Deforestation Patterns** – Adam Voiland – NASA Earth Observatory – https://earthobservatory.nasa.gov – Accessed: October 2020
[94] **What Uncontacted Tribes Want & What We Can Do for the Amazon** – José Carlos Meirelles — https://www.youtube.com/watch?v=QB2YPHAWQcM – Accessed: November 2020
[95] **Uncontacted tribes: the threats** – Survival International – https://www.survivalinternational.org/ – Accessed: November 2020

single man alone resides. No one alive knows his name. He is the last of his tribe.[96] The Brazilian government gave up trying to make contact with him because he shot arrows at anyone who came too close.[97] When his solitary life comes to an end, his tribe will be no more and the patch of land he is occupying will surely be consumed by agriculture. He is not alone in one forlorn respect: Other uncontacted tribes are also losing their land.[98]

Shrinking Wildfires: Shrinking Wilderness

The surprising 24.3% *decrease* in global wildfire burn areas over the 18-year period ending in 2015 (based on satellite data) is a reflection of the scale of the transformation of wilderness into farmland that is occurring on our Earth.[99] The expansion of farmland at the expense of wilderness will be further developed in the next section "Farmland: Wildfire Boundary".

Whether North American bison, African elephants and the uncontacted tribes of the Amazon thrive or perish will depend entirely on whether they have enough wild land on which to live. Contrary to widely held beliefs, habitat loss – not global warming – is the greatest threat to wildlife.[100]

In the sections "Farmland: Wildfire Boundary" and "From Fire: Human Development" we will see how the human use of fire has increased the efficiency of farming. Efficient farming requires less

---

[96] **Last survivor: The story of the 'world's loneliest man'** – Vicky Baker – BBC News – 20 July 2018

[97] **This man is the last of his tribe. Let him live in peace** – Fiona Watson – Survival International – survivalinternational.org – Accessed: November 2020

[98] **Some Isolated Tribes in the Amazon Are Initiating Contact** – Scott Wallace – National Geographic – 13 August 2013; and
First Contact: Lost Tribe of the Amazon – UK Chanel 4 Documentary – 23 February 2016

[99] **A human-driven decline in global burned area** – N. Andela, D. C. Morton, L. Giglio, Y. Chen, G. R. van der Werf, P. S. Kasibhatla, R. S. DeFries, G. J. Collatz, S. Hantson, S. Kloster, D. Bachelet, M. Forrest, G. Lasslop, F. Li, S. Mangeon, J. R. Melton, C. Yue, J. T. Randerson – Science – 30 June 2017

[100] **Nature's Dangerous Decline 'Unprecedented' Species Extinction Rates 'Accelerating'** – United Nations, Intergovernmental Science-Policy on Biodiversity and Ecosystem Services – 6 May 2019

land to produce the same amount of food, which increases the land available for wildlife.

We have assessed changing wildfire regimes and changes to the Earth's wilderness in several continents over several centuries. In the next section we will assess the core driver of those changes, a need for more of one thing: farmland.

4. Farmland: Wildfire Boundary

Wildfire dynamics change abruptly at the boundary between farmland and wilderness, but what are the root causes driving the expansion of farmland and why is this expansion currently most pronounced in the Tropics? Those questions will be answered in this section.

In this section we will also gain an understanding of the importance of agricultural yields. In the section "From Fire: Human Development" we will gain an appreciation of the role that the human use of fire has played, and continues to play, in increasing agricultural yields.

Farmland: the Basis of Civilization

Since the acquisition of language some 80,000 years ago, the most significant development in our human trajectory has been the development of agriculture – farming. The most significant and clearly evidenced cause of the Agricultural Revolution was the end of the most recent ice age. If we accept that humans have been present in our current anatomical form for the last 300,000 years, we would have spent the vast majority of our existence struggling to survive against cold and ice. This is known because the last 300,000 years have generally been glacial periods. Prior to about 18,000 years ago, global temperatures, as estimated by the US National Oceanic and Atmospheric Administration, were as much as 7.5° Celsius (13.5° Fahrenheit) colder than they are today.[101]

Glaciers occur when snow does not melt away in the summer over many years, which allows it to accumulate, compact and turn into ice. Over the years, as more snow accumulates and is transformed into ice, glaciers can thicken into massive ice sheets several kilometers high.

---

[101] **Glacial-Interglacial Cycles** – National Oceanic and Atmospheric Administration – https://www.ncdc.noaa.gov – Accessed: July 2020

During the last glacial period, what is now the city of New York was entirely covered in ice sheets. Those ice sheets are estimated to have been higher than the skyscrapers that now give that city its skyline.[102]

Recent research has provided incredibly granular knowledge of the ice-sheet cover over Britain and Ireland. The maximum extent of the most recent ice sheets over Britain and Ireland occurred some 22,000 years ago. At that time, Ireland and Scotland were entirely covered by ice and Wales was largely covered in ice too. Ice sheets in England came as far south as to almost, but not quite, cover the locations of the current cities of Leeds, Manchester and Sheffield.[103] However, around 15,000 years ago global temperatures began to rise. As they did, great areas of the Earth that had been covered by ice were transformed into fertile forests and grasslands. Britain was generally free of ice from 11,700 years ago.

The beginning of the Agricultural Revolution could not have been triggered only by a warmer climate because there were two prior interglacial periods over the last 300,000 years during which estimated temperatures were actually higher than they are today (Figure 8), but there is no evidence of a shift to farming during any of the prior interglacial periods.[104]

---

[102] **How the Ice Age Shaped New York** – William J. Broad – New York Times – 5 June 2018

[103] **BRITICE Glacial Map, version 2: a map and GIS database of glacial landforms of the last British–Irish Ice Sheet** – Chris D. Clark, Jeremy C. Ely, Sarah L. Greenwood, Anna L. C. Hughes, Robert Meehan, Iestyn D. Barr, Mark D. Bateman, Tom Bradwell, Jenny Doole, David J. A. Evans, Colm J. Jordan, Xavier Monteys, Xavier M. Pellicer, Michael Sheehy – BOREAS An International Journal of Quaternary Research – 29 August 2017

[104] **Glacial-Interglacial Cycles** – National Oceanic and Atmospheric Administration – https://www.ncdc.noaa.gov – Accessed: July 2020; and
Northern Hemisphere forcing of climatic cycles in Antarctica over the past 360,000 years – Kenji Kawamura, Frédéric Parrenin, Lorraine Lisiecki, Ryu Uemura, Françoise Vimeux, Jeffrey P. Severinghaus, Manuel A. Hutterli, Takakiyo Nakazawa, Shuji Aoki, Jean Jouzel, Maureen E. Raymo, Koji Matsumoto, Hisakazu Nakata, Hideaki Motoyama, Shuji Fujita, Kumiko Goto-Azuma, Yoshiyuki Fujii & Okitsugu Watanabe – Nature – 23 August 2007

Glacial and interglacial periods — Figure 8

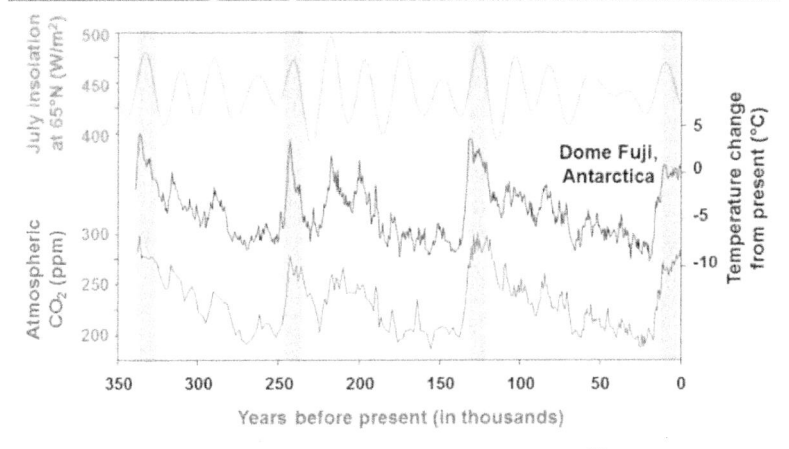

Image Credit: US National Oceanic and Atmospheric Administration[105]

As we have seen, we acquired language (the cognitive ability to code ideas in our minds) some 80,000 years ago. The first interglacial period to have occurred since acquiring language is the one in which we are currently living. Having acquired language, all that was needed to kick-start the Agricultural Revolution and the emergence of human civilizations was a change in global temperatures and the corresponding retreat of ice sheets.

Prior to the end of the last ice age, humans lived in small nomadic hunter-gatherer communities. 23,000 years ago, the European-wide human population was less than 200,000 by some estimates.[106]

From that statistic alone we know that survival was not to be taken for granted during the last ice age. The combination of less food and considerably colder temperatures would have made life extremely difficult indeed.

As the ice sheets retreated, humans were thriving. Food sources became more plentiful and our hunting skills were honed to make the

---

[105] **Glacial-Interglacial Cycles** – National Oceanic and Atmospheric Administration – https://www.ncdc.noaa.gov – Accessed: July 2020
[106] **Human population dynamics in Europe over the Last Glacial Maximum** – Miikka Tallavaara, Miska Luoto, Natalia Korhonen, Heikki Järvinen and Heikki Seppä – Proceedings of the National Academy of Sciences of the United States of America – 7 July 2015

most of the opportunity. Human population levels increased during this period. However, at some point, the trend of rising human population levels created an existential threat for many humans: There was simply not enough food to support a continuously growing population.

Land is limited and hunting and gathering requires a tremendous amount of land to support a limited number of people. By some estimates, our Earth can support a population of only 10 million hunter-gatherers.[107]

From the time that humans acquired language to 1,000 years ago, 177 large mammals went extinct in what is known as the global Megafaunal Extinction.[108] This extinction event started well before the transition from hunter-gatherers to farmers began; however, it would surely have been intensified by rising populations of hungry hunter-gatherers.[109] As a result of these extinctions, hunter-gatherers lost sources of food.

With insufficient food to support their growing populations and with sources of food diminishing, many humans were at risk of perishing.

From experimental origins and across the world, societies gave up their nomadic ways to become sedentary agriculturalists because agriculture provided a means of ensuring that growing human populations would have enough food. There is no evidence to suggest that the quality of food or the lifestyles of early farmers was better than it had been for hunter-gatherers. However, farming reduced the risks of hunger and starvation. Agriculture supported ever-increasing human populations making a return to nomadic ways impractical.

---

[107] **Hunter-gatherer populations inform modern ecology** – Joseph R. Burger and Trevor S. Fristoe – Proceedings of the National Academy of Sciences of the United States of America – 6 February 2016

[108] **Global late Quaternary megafauna extinctions linked to humans, not climate change** – Christopher Sandom, Søren Faurby, Brody Sandel and Jens-Christian Svenning – Proceedings of the Royal Society – 22 July 2014; and
Extinction of large mammals in the Late Quaternary Ice Age – Adrian Lister – Natural History Museum – https://www.nhm.ac.uk – Accessed: February 2021

[109] **Population reconstructions for humans and megafauna suggest mixed causes for North American Pleistocene extinctions** – Jack M. Broughton and Elic M. Weitzel – Nature Communications – 21 December 2018

The earliest farming communities emerged in:[110]

i) the eastern Mediterranean, from which area farming extended quickly to adjacent areas with fertile land;

ii) Central Mexico; and

iii) the middle Yangtze River region in China.

The archaeological evidence from the eastern Mediterranean region provides the best record of the early transition of hunter-gatherers into sedentary farmers. The transition to a sedentary lifestyle was a slow transition which started with the harvesting of wild cereals. Cereal grains keep for a long time, which is critically advantageous. The possibility of storing large amounts of grain would have been one important reason to have abandoned an entirely nomadic existence. The first settled people of the area, the Early Natufians, lived 12,500 to 11,000 years ago and they built stone houses that had a resemblance to modern houses.[111]

The Agricultural Revolution consisted of applying human ingenuity to increase food production from a limited area of land. For example, the primitive Natufian sickles increased yields by reducing waste.

The first farmers intended only to produce enough food to sustain their communities, but soon they were creating a surplus of food. This allowed people to engage in other specialist activities and to move to newly emerging towns. From these towns and from the freedom provided by a surplus of food rose social structures, commerce, armies, professional clergy, specialization of labor and governments. Indeed, the great civilizations of antiquity all rose up as a product of the Agricultural Revolution – as have our own.

The mathematical logic of agricultural yields and land use can be shown by example. In 2017, the global average yield for wheat fields was 3.46 metric tons per hectare and for rice that figure was 3.10 metric tons per hectare. For reference, a hectare is an area of land equal to 100 meters by 100 meters or roughly 2.5 acres. In the same year, the world consumed 761.3 million metric tons of wheat and

---

[110] **The Natufian Culture in the Levant, Threshold to the Origins of Agriculture** – Ofer Bar-Yosef – Evolutionary Anthropology – 7 December 1998

[111] **The Natufian Culture in the Levant, Threshold to the Origins of Agriculture** – Ofer Bar-Yosef – Evolutionary Anthropology – 7 December 1998

512.4 million metric tons of rice.[112] For any particular food, dividing the amount of food we need by the agricultural yield for that food indicates how much land we need to satisfy our requirements. To satisfy our needs in wheat, we need 2.2 million square kilometers (849 thousand square miles) of farmland – equivalent to 3.2 times the surface area of the state of Texas, USA. To satisfy our needs in rice, we need 1.7 million square kilometers (638 thousand square miles) of farmland – equivalent to 2.4 times the surface area of the state of Texas, USA.

The human population has grown to 7.7 billion and is expected to grow by a further 42% by the year 2100.[113] If agricultural yields do not increase correspondingly by that time, in order to grow enough food, we will need more farmland. In effect, agricultural yields must increase over the next century to avert the destruction of wildlife on a colossal scale.

Farmland: Taken from Wilderness

63% of the Earth's land surface is suitable for the growth of forests or grasslands. Of that land, humans have permanently settled 56%.[114]

We use 93% of the land we have permanently settled for agriculture and the remaining 7% for urban and infrastructure related purposes. 30% of agricultural land is used for crops and the remaining 70% is used as pastureland for grazing livestock.[115]

By some estimates, three-quarters of the Earth's land environment has been significantly altered by human actions.[116]

---

[112] **Agricultural Outlook 2019-2028** – OECD/Food and Agricultural Organization UN – https://stats.oecd.org – Accessed: March 2020
[113] **World Population Prospects 2019** – United Nations – Department of Economic and Social Affairs
[114] **Assessing Global Land Use** – Stefan Bringezu, Helmut Schütz, Walter Pengue, Meghan O'Brien, Fernando Garcia, Ralph Sims, Robert W. Howarth, Lea Kauppi, Mark Swilling and Jeffrey Herrick – United Nations Environment Programme – 2013
[115] **Assessing Global Land Use** – Stefan Bringezu, Helmut Schütz, Walter Pengue, Meghan O'Brien, Fernando Garcia, Ralph Sims, Robert W. Howarth, Lea Kauppi, Mark Swilling and Jeffrey Herrick – United Nations Environment Programme – 2013
[116] **Nature's Dangerous Decline 'Unprecedented' Species Extinction Rates 'Accelerating'** – United Nations, Intergovernmental Science-Policy on Biodiversity and Ecosystem Services – 6 May 2019

Since the beginning of the Agricultural Revolution some 12,500 years ago, the expansion of farmland and urbanization has displaced wildlife.

Habitat loss has been, is, and will be, the principal threat to terrestrial ecosystems – wildlife – because without land wildlife cannot exist.

Conservationists have long known that habitat loss is the single greatest threat to wildlife. The World Wildlife Fund and the National Wildlife Federation of the United States have both drawn attention to this issue by stating that habitat loss is the number-one threat to wildlife.[117]

In 2019, for the first time ever, the United Nations sponsored a global intergovernmental project to assess and rank risks to wildlife. That research group determined that globally the single most important risk to wildlife is habitat loss.[118]

For reference, according to that research project, the second greatest threat to wildlife is the direct exploitation of animals and the third greatest threat to wildlife is global warming.

Deforestation for the purposes of creating farmland causes extreme transformations of landscapes. Since the launch of NASA's Terra satellite, it has been possible to track changes to tree cover on a global basis. Bearing in mind that satellite data always provides completely surprising results, let us take a look at the rates of deforestation globally.

Globally, and contrary to prevailing expectations, the Earth's forested surface area has actually *increased* at a rate of 175 square kilometers (68 square miles) *per day* over the 35 years to 2016, net of losses, based on satellite data. In total, over that period the forested surface area of the Earth has increased by 7% or 2.24 million square

---

[117] **"Habitat Loss Poses the Greatest Threat to Species" Habitat Loss** – World Wildlife Fund – https://wwf.panda.org/ – Accessed: July 2020; and
Threats to Wildlife – The National Wildlife Federation – https://www.nwf.org/ – Accessed: July 2020
[118] **Nature's Dangerous Decline 'Unprecedented' Species Extinction Rates 'Accelerating'** – United Nations, Intergovernmental Science-Policy on Biodiversity and Ecosystem Service – 6 May 2019

kilometers (864 thousand square miles), net of losses.[119] In contrast to widely held perceptions, in general, globally, forests are thriving. This will be assessed in more detail in the section "From Fire: Carbon Dioxide Fertilization".

In contrast to the global trend, in the Tropics forests are in decline.

Based on satellite data, deforestation has been most pronounced in the southern Tropics. Brazil recorded the highest rate of tree cover loss over the period (1982-2016) with a loss of 385 thousand square kilometers (149 thousand square miles), representing a percentage decline of 8% over the period.

The satellite data measures tree cover and does not distinguish between a forest and a tree plantation. In the Tropics deforestation is occurring partly to cultivate fruit that grows in trees, such as palm fruits. Therefore, the satellite data under-represents deforestation in the Tropics.

Using African data that has been collected by on-the-ground surveys and sorting national data by geography shows that in the decade ending in 2010 the African Tropics lost 326 thousand square kilometers (126 thousand square miles) of forest, representing a loss of 4.9% of forested land over that decade.[120]

The key points are as follows:

i) Habitat loss represents *by far* the single most important threat to wildlife globally.

ii) Wildlife has been losing habitat due to the expansion of the Agricultural Revolution for the last 12,500 years as people have settled land, principally for agricultural and urban uses.

iii) Today, habitat loss is concentrated in the Tropics.

---

[119] **Global land change from 1982 to 2016** – Xiao-Peng Song, Matthew C. Hansen, Stephen V. Stehman, Peter V. Potapov, Alexandra Tyukavina, Eric F. Vermote & John R. Townshend – Nature – 8 August 2018
[120] **Forests, Trees, and Woodlands in Africa** – Africa Region, World Bank – 11 October 2012

Farmland: to End Hunger

Let us turn our attention to the root cause of the Agricultural Revolution: hunger.

Having grown up in the 1980s, I can remember that in that decade hunger was a major global concern. Surely, hunger is not being talked about as much today because it has been eradicated; let us take a look at the scorecard.

Based on data from the United Nations, in the 1970s the number of people suffering from malnourishment globally was reduced by 37 million people. By the end of the 1980s the number of people suffering from malnourishment had fallen by a further 100 million.[121] However, in the three decades since the 1980s to the present no more than 3 million people have left hunger globally based on data from the same source.[122] Today, 820 million people continue to suffer from hunger.[123] In essence, after two decades of profound improvement in the alleviation of malnourishment, there has been a period of three whole decades during which there has been no substantive improvement in malnourishment globally.

In a report entitled "The Neglected Crisis of Undernutrition: Evidence for Action" the United Kingdom's Department for International Development indicated that "nutrition fundamentally determines life chances and people's ability to convert opportunities into outcomes."[124] That puts into focus the World Health Organization's estimate that 144 million children under the age of five will be stunted, implying their physical and mental development

---

[121] **Undernourishment around the world, Counting the hungry: trends in the developing world and countries in transition; The State of Food Insecurity in the World 2006** – Food and Agriculture Organisation of the United Nations – 2006
[122] **The State of Food Security and Nutrition in the World 2019** – Food and Agriculture Organization of the United Nations, International Fund for Agricultural Development, UNICEF, World Food Programme and World Health Organization – 2019
[123] **The State of Food Security and Nutrition in the World 2019** – Food and Agriculture Organization of the United Nations, International Fund for Agricultural Development, UNICEF, World Food Programme and World Health Organization – 2019
[124] **The Neglected Crisis of Undernutrition: Evidence for action** – UKAID, UK Department of International Development – 2009

will be damaged permanently due to a lack of food.[125] That figure represents 21.3% of young children globally.[126] Today, 45% of child deaths globally are associated with malnutrition according to the World Health Organization.[127]

The United Nations also estimated that, in addition to the people who are hungry today, there are 2 billion people who are experiencing moderate to severe food insecurity. This implies that a great many people are at risk of being unable to assure their food in sufficient quantity and quality.

820 million hungry. 144 million stunted children. 2 billion experiencing food insecurity. No improvement in three decades.

That is what you call one utterly repulsive scorecard.

In remarkable contradiction to that assessment, many organizations that track hunger indicate that decade-over-decade hunger has been in decline.[128] What is going on?

As the number of people suffering from hunger has remained unchanged for the last three decades, the global population has grown. As a result, the percentage of the global population suffering from hunger has fallen. On that basis, many organizations have communicated that hunger is in decline. Further confounding public perception, there are multiple indicators that are used to estimate the extent of global malnourishment.[129]

The key point to retain is that, despite many statements to the contrary, since the 1980s the number of our fellow humans suffering from hunger globally has not fallen meaningfully and today,

---

[125] **World Health Statistics 2020** – World Health Organization: monitoring health for the SDGs, sustainable development goals – 2020

[126] **Child Stunting** – World Health Organization – https://www.who.int – Accessed: March 2021

[127] **Malnutrition** – World Health Organization – https://www.who.int – Accessed: March 2021

[128] **The State of Food Security and Nutrition in the World 2019** – Food and Agriculture Organization of the United Nations, International Fund for Agricultural Development, UNICEF, World Food Programme and World Health Organization – 2019

[129] **Hunger and Food Insecurity** – United Nations Food and Agriculture Organization – http://www.fao.org/ – Accessed May 2021

according to the United Nations, "More than 820 million people do not have enough food to eat."[130]

Of most alarm, the United Nations has indicated that it expects hunger to worsen significantly in the current decade as measured by the number of people suffering from hunger and by the percentage of the global population suffering from hunger.[131]

In contrast to the widely held perception that hunger globally is not an issue of primary concern, it represents the greatest humanitarian crisis of modern times. Critically, hunger is also a root cause of many other global tragedies, inclusive of the large-scale destruction of wildlife habitat, and correspondingly, of wildlife.

Let us now turn our attention to where global hunger and extreme poverty are geographically concentrated.

Two Finnish professors had the idea of plotting measures of human development against lines of latitude from the North Pole, through the Equator, to the South Pole. Based on their data, extreme poverty is distinctly concentrated in the Tropics.[132]

Currently, 670 million people living in the Tropics are estimated to be suffering from extreme poverty.[133] 85% of the poorest people in the world live in the Tropics.[134] It is projected that as many as 67% of the world's children under the age of 15 will be living in the Tropics by 2050.[135]

---

[130] **The State of Food Security and Nutrition in the World 2019** – Food and Agriculture Organization of the United Nations – http://www.fao.org/state-of-food-security-nutrition – Accessed: May 2021

[131] **Hunger and Food Insecurity** – United Nations Food and Agriculture Organization – http://www.fao.org/ – Accessed May 2021

[132] **The world by latitudes: A global analysis of human population, development level and environment across the north–south axis over the past half century** – Matti Kumma and Olli Varis – Applied Geography – April 2011

[133] **State of the Tropics 2020 Report** – Sandra Harding, Ann Penny, Shelley Templeman, Madeline McKenzie, Daniela Tello Toral and Erin Hunt – State of the Tropics – 2020

[134] **State of the Tropics 2020 Report** – Sandra Harding, Ann Penny, Shelley Templeman, Madeline McKenzie, Daniela Tello Toral and Erin Hunt – State of the Tropics – 2020

[135] **Explore the Data** – State of the Tropics – https://www.jcu.edu.au/state-of-the-tropics/data – Accessed May 2021

Farmland: a Cornerstone

Fire has been essential for human existence since we emerged as a species some 300,000 years ago. Farming has been essential for our existence since we became reliant on it for our food, starting some 12,500 years ago. Fire and agriculture are the two cornerstones of human development.

Since the time of the first farmers, the Natufians, people have turned wilderness into farmland to address hunger and desperation. Today, hunger and desperation are concentrated in the Tropics. As a result, the expansion of farmland and the corresponding destruction of wildlife habitat are most pronounced in the Tropics.

Increasing agricultural yields is the most time proven means of increasing prosperity,[136] reducing hunger[137] and increasing the availability of land for wildlife.

Tanzania, the host country of the Serengeti National Park, provides a reference: Tanzania's main crop is corn[138] and the average corn yield in Tanzania is 1.5 metric tons per hectare (21 bushels per acre).[139] Although Tanzania has fertile land, its corn yield is low by international standards. The evolution of Tanzania's corn yield over the coming decades will be determinant for i) the future prosperity of the people of Tanzania who are currently 58 times poorer on average than Americans[140] and ii) the availability of wildlife habitat for the wild animals that live in that country.

Many of the risks that are widely perceived to be detrimental to wildlife – inclusive of global warming – are insignificant relative to the risk of habitat loss: The African elephant population has fallen by a factor of 20 times since 1930 because African elephants have lost

---

[136] **Ending Extreme Poverty** – Interview of Ana Revenga by Amy Frykholm – 8 June 2016 – The World Bank (as first published by the Christian Century)
[137] **Ending Extreme Poverty** – Interview of Ana Revenga by Amy Frykholm – 8 June 2016 – The World Bank (as first published by the Christian Century)
[138] **The Maize Value Chain in Tanzania** – Food and Agricultural Organization of the United Nation – R. Trevor Wilson and J. Lewis – 2015
[139] **Tanzania, Grain and Feed Annual** – US Department of Agriculture – Ben Mtaki – 22 April 2020
[140] **GDP per Capita, 2019** – World Bank – https://data.worldbank.org – Accessed: November 2020

their habitat – their land – not because of global warming. Positively, when wildlands are recovered and given back to wildlife, wildlife populations recover. If Tanzania's corn yield was comparable to that of the United States (10.8 metric tons per hectare; 172 bushels per acre),[141] it would create an extraordinary amount of stomping room for African elephants. This is because only 12% of the land in Tanzania being used to grow corn would be required for that purpose. That would free 88% of the land currently being used to grow corn in Tanzania for other purposes, including providing wildlife with habitat.

But what does this have to do with the human use of fire?

Everything.

Fire has been instrumental in increasing agricultural yields from the Bronze Age to the present. This will be developed in the section "From Fire: Human Development".

The most important impact on wildlife resulting from the replacement of fire from burning coal, oil and natural gas with alternatives will relate to how that change affects agricultural yields. Whether we need more land to grow food, or less, is the factor that will determine whether wildlife thrives or perishes.

The frontier between farmland and wilderness represents the boundary between the wildfire regimes of nomadic and sedentary people. Having looked at farmland, the Agricultural Revolution and the importance of agricultural yields, let us now turn our attention to the technologies derived from fire.

---

[141] **Corn and soybean production up in 2020, USDA Reports** – US Department of Agriculture – 12 January 2021

Part 2: From Fire: Human Development

1. Fire Starting: Our First Tech

Today, we give very little thought to how fires are lit. Spark plugs within internal combustion engines can spark the ignition of thousands of fires every minute without us even being aware of it. When we turn on the lights, we rarely realize that a fire has been lit on the other side of the electricity grid to provide the energy for that to happen. We will see in the section "From Fire: Energy" that fire provides more than half of electrical energy globally.[142]

In contrast to our current indifference to fire, fire would have been at the forefront of our thoughts for most of our existence as a species. We think of pre-agricultural humans as nomadic hunter-gatherers; however, it might be more accurate to describe them as fire starters.

In many cases, hunting would have been the easy part of providing food, and getting a fire lit would have been the principal challenge. One can imagine the countless number of cold, snowy nights the eyes of small communities would have strained in the hopes of seeing flames erupt from the work of fire starters. In many instances, survival itself would have depended on the outcome.

We do not have definitive answers as to how we first developed our most important and fundamental technology, the lighting of fire.[143] We do know that techniques to light fires using only natural materials depend principally on the materials that are available locally.

In areas where pyrite is available, that material is an obvious choice for creating sparks that can be used to start fires. Pyrite has special properties as its name would imply, "pyr" means fire in Greek. Pyrite actually consists of iron sulfide (FeS_2). When pyrite is struck by a

---

[142] **Electricity Information Overview 2020 (data for 2018)** – International Energy Agency – July 2020
[143] **A Spark in the Dark: Shedding Light on the Origins of Fire-making** – Andrew Sorensen – leidenarchaeologyblog – 10 September 2019 – https://leidenarchaeologyblog.nl – Accessed: July 2020;
The discovery of fire by humans: a long and convoluted process – J. A. J. Gowlett – Philosophical Transactions of the Royal Society – 5 June 2016; and
Who Started the First Fire? – Dennis Sandgathe and Harold L. Dibble – Sapiens – 26 January 2017 – https://www.sapiens.org – Accessed: July 2020

hard surface, part of it can chip off and expose a new surface of iron sulfide to the oxygen in the air, causing a fast exothermic chemical reaction – also called a brief fire. The resulting heat can cause the chips to momentarily glow red, creating a spark.

Flint is used to strike pyrite because it is hard and can be crafted to have sharp edges.

From the time of the Iron Age, steel has replaced pyrite for the purposes of creating sparks. Steel will be discussed in the section "From Fire: the Iron Age".

Making fire with sticks is significantly more difficult than doing so using pyrite and flint.

Although there are many techniques to create fire using sticks, they all involve the creation of friction to i) heat wood in order to dry it out and thereby lower its ignition temperature, ii) create fine wood dust and iii) create sufficient heat to ignite that wood dust.

With the mastery of fire, humans were able to develop materials that would serve as the building blocks of our civilizations. These materials and the story of their acquisition is the subject of the next section.

2. From Fire: Materials

Prehistory, classically, is divided into the Stone Age, the Bronze Age and the Iron Age. However, no material has served as a more constructive building block for human development than the brick, or more specifically, the fired brick. It was on bricks that we first developed the greatest legacy of the ancient world: writing. The commonality between bricks, bronze and iron is that they are all the product of fire. Let us take a better look at these materials and how they shaped the course of humanity.

From Fire: Bricks, Glass and Cement

The Venus of Dolni Vestonice, shown in Figure 9, is one of the oldest objects of representational art to have been discovered. It was unearthed at the Paleolithic site of Brno in the Czech Republic in 1925. It is dated to around 29,000 years ago.[144] For reference, the Lion Man of Germany's Hohlenstein-Stadel Cave was discovered in 1939 and dated to about 40,000 years ago.[145] Although both objects are considered to be masterpieces of the ice age, the reality is that they are the only objects of their kind to have survived. Their particularity is that they have endured.

The Lion Man, which was carved in mammoth ivory, is actually an assemblage of more than 300 broken fragments.[146] The object that we can see at the Museum in Ulm is therefore quite unlike the object that was created 40,000 years ago. Due to the condition of the Lion Man there is even debate as to whether or not the Lion Man was intended to be male.[147] The Venus of Dolni Vestonice too was broken, but only in two pieces and its surface is in excellent condition. The Venus of Dolni Vestonice is the oldest known

---

[144] **The Adiposity Paradox in the Middle Danubian Gravettian** – Erik Trinkaus – Anthropologie – 2005
[145] **The Lion Man** – Museum Ulm – https://museumulm.de/ – Accessed: July 2020
[146] **The Lion Man** – http://www.loewenmensch.de/lion_man.html – Accessed: July 2020
[147] **Is the Lion Man a Woman?** – Von Matthias Schulz – Spiegel International – 9 December 2011

ceramic. It is due to its properties as a ceramic that it has survived in excellent condition for 29,000 years.

From fire: 29,000-year-old Venus of Dolni Vestonice *Figure 9*

Photo Credit: Author

Ceramics consist of earthen clay materials that have been fired. During the firing process, through the application of heat, provided by fire, clay is partially vitrified – partially turned into glass. Once vitrified, ceramics are both strong and waterproof.

Bricks are simply ceramic blocks that are produced by heating clay using fire.

The Agricultural Revolution spread from the first Natufians in the eastern Mediterranean to adjacent areas with fertile land. The great rivers of that region, namely, the Nile, the Euphrates and the Tigris provide water that makes nearby land fertile. Due to the crescent shape of the region's fertile land, it is known as the Fertile Crescent.

As agricultural yields increased in the Fertile Crescent, increased food supplies allowed the first cities to emerge in the area. Erbil, in the Kurdistan region of Iraq, is possibly the oldest continuously occupied human settlement on Earth, with urban life there starting in

6,000 BC.[148] Erbil's urban layout is representative of how the first cities were built. Each city had a central fortress, or citadel, built on raised land.[149] Surrounding the raised land there was a wide ring of land in which people would live. Surrounding that, a circular fortification wall was built.

Throughout the Fertile Crescent mud bricks were used for construction. Fortresses, fortification walls and homes were all built with mud bricks. Mud bricks are just that – bricks made of mud that have not been fired.

The Sumer civilization emerged in southern Mesopotamia around 4,500 BC.[150] In that area the Euphrates and the Tigris rivers merge into marshy wetlands before emptying into the Persian Gulf. Today, Basrah, Iraq, is a notable city in the area of ancient Sumer.

The Sumerians were disadvantaged because the wet and marshy environment in which they lived would quickly deteriorate mud brick constructions. By firing mud bricks to create vitrified ceramic bricks the Sumerians found a solution to their challenge, a solution that would forever transform construction and architecture.

Fired bricks are able to withstand the elements and they are considerably stronger than mud bricks. For these reasons, the use of fired bricks spread throughout the Fertile Crescent. Fired bricks allowed for the construction of greater fortresses and fortifications, which gave rise to greater city states that ruled over larger territories. The Great Ziggurat of Ur, which was built in brick around 2,100 BC in Southern Sumer, is representative of these developments. That ziggurat as partially reconstructed in the 1980s is shown in Figure 10. It is an impressive example of how materials and technology shape social and political structures.

---

[148] **Erbil History** – Salahaddin University-Erbil – https://su.edu.krd/about/history/erbil – Accessed: July 2020;
History on a Hill – Kasha Patel – NASA Earth Observatory – https://Earthobservatory.nasa.gov – Accessed: July 2020; and
Erbil Revealed – Andrew Lawler – Archaeology Today – September/October 2014
[149] **Erbil Revealed** – Andrew Lawler – Archaeology Today – September/October 2014
[150] **The Sumerians** – Samuel Noah Kramer – The University of Chicago Press – 1963

From fired bricks: the Great Ziggurat of Ur *Figure 10*

Photo Credit: US Air Force, Staff Sergeant Christopher Marasky

Sumerians were an enterprising, trading people. They would exchange tokens as part of commercial transactions. For example, a token might mean "I owe you a goat" or "I will pay you one shekel of silver." To avoid fraud and to keep their transactions organized, Sumerians placed the tokens related to each important transaction into a specific clay ball. Each clay ball would then be sealed and fired with the tokens inside it. If evidence of a transaction was required, the clay ball would be broken open to reveal the pertinent tokens. As reminders of the contents within the balls, Sumerians began pressing the tokens onto the exterior of the clay balls before they were sealed and fired. Over time, the impressions themselves became the primary record of significance. Ultimately, the Sumerians realized that a permanent record of their transactions could simply be made directly onto flat clay tablets – bricks of a different form.[151]

In this way, over the period from about 3,500 BC to about 3,100 BC the symbols of accounting units – tokens – evolved into one of the most complete writing systems the world has known, cuneiform writing – the first form of writing. For reference, the earliest Egyptian hieroglyphs started emerging only near the end of the noted period.

Throughout history, bricks have been used for the same reasons as Sumerians recorded writing on them: They are simple to make, they are strong and they last. The Romans industrialized brick production and spread the use of brick making. They were not the only ancient civilization to have made bricks on an industrial scale: The largest

---

[151] **The Evolution of Writing** – Denise Schmandt-Besserat – International Encyclopaedia of Social and Behavioural Sciences – Elsevier – 2014

man-made structure on Earth, the Great Wall of China, is largely made of fired bricks.

As we have seen, heat is required to induce the chemical reactions required to create bricks. An idealized brick firing process would involve heating bricks for about three and a half days, of which 10 hours would involve heat over 1,000° Celsius (1,832° Fahrenheit).[152]

The creation of other building materials such as glass and cement also require tremendous amounts of heat. The heat to produce these materials has historically been provided by fire. Glass production was first developed in Mesopotamia around 2,000 BC by melting sand at temperatures of 1,700° Celsius (3,090° Fahrenheit).[153] Cement was perfected by the Romans, which allowed them to build in the 2nd century AD the dome of the Pantheon – still the world's largest unreinforced cement dome. To produce cement, limestone and clay must be heated to approximately 1,450° Celsius (2,640° Fahrenheit).[154]

Turning our focus of attention to the present, fire, or more specifically heat, remains essential to the production of bricks, cement and glass. At present, the cost of heat as a percentage of the total production costs for bricks, cement and glass are respectively estimated to be 30-35%, 30-40% and 15%.[155] The key point is that the cost of heat represents a very large component of the costs of producing these basic materials.

---

[152] **The UK Clay Brick Making Process** – Brick Development Association – www.brick.org – March 2017
[153] **The Origins of Glassmaking** – Corning Museum of Glass – https://www.cmog.org/article/origins-glassmaking – Accessed: August 2020
[154] **Climate change: The massive CO2 emitter you may not know about** – Lucy Rogers – BBC News – 17 December 2018
[155] **Final Report: For a study on the composition and drivers of energy prices and costs in energy intensive industries: The case of the ceramics industry, bricks and roof tiles** – Christian Egenhofer, Lorna Schrefler, Fabio Genoese, Julian Wieczorkiewicz, Lorenzo Colantoni, Wijnand Stoefs, Jacopo Timini – Center for European Policy Studies – 13 January 2014;
Improving thermal and electric energy efficiency at cement plants: intranational best practice – International Finance Corporation, World Bank Group – 2017; and
Glass Industry of the Future, Energy and Environmental Profile of the US Glass Industry – Joan L. Pellegrino, Lou Sousa, Elliott Levine, Michael Greenman, C. Philip Ross, Jim Shell, Marv Gridley, Dan Wishnick and Derek J. McCracken – Office of Industrial Technologies, US Department of Energy – April 2002

Remaining in the present, currently, 1,500 billion fired bricks are produced per year.[156] That equates to 195 bricks per person living on Earth per year. Current country rankings for brick and cement production are provided in Table 1.

Brick & cement production by country *Table 1*

Brick Production		Cement Production	
Country	**%**	**Country**	**%**
China	67%	China	54%
India	16%	India	7%
Pakistan	3%	Vietnam	2%
Other	14%	Other	37%
Total	**100%**	**Total**	**100%**

Sources[157]

Fire has been used to create basic building materials for millennia. Today, fire is being used to create bricks and cement principally in developing countries that are relatively poor and experiencing extraordinary urbanization pressures. From this perspective, the benefits of fire for producing bricks and cement are *by far* the most pronounced in relatively poor developing countries such as China and India.

Unlike in developing countries, bricks in wealthy countries have lost their place as a material that is integral to construction. The Empire State Building of New York was completed in 1931. Its height of 381 meters (1,250 feet) made it the tallest building in the world at that time. 10 million bricks form the exterior façade of that building.[158] However, its structure is not supported by bricks, which are there merely for aesthetic decoration. The Empire State Building is supported structurally by something else entirely: steel. Let us now turn our attention to metallurgy, the development of which brought

---

[156] **Emissions from South Asian Brick Production & Potential Climate Impact** – Ellen Baum – Climate and Health Research Network, presentation – 11 March 2015 – Accessed at https://cdn.cseindia.org/ (August 2020)

[157] **Emissions from South Asian Brick Production & Potential Climate Impact** – Ellen Baum – Climate and Health Research Network, presentation – 11 March 2015 – Accessed at https://cdn.cseindia.org/ (August 2020); and
Cement Data Sheet, Mineral Commodity Summaries 2020 – US Geological Survey – https://pubs.usgs.gov – Accessed: August 2020

[158] **Empire State Building Fact Sheet** – Empire State Realty Trust – https://www.esbnyc.com – Accessed: February 2021

an end to the Stone Age and ultimately allowed humanity to build into the skies.

From Fire: the Bronze Age

Bricks, cement and glass are the products of fire. We will now see that bronze and metallurgy are also the products of fire.

Bronze is an alloy resulting from the combination of copper and tin. The story of bronze begins with the story of copper. The use of copper represents the origins of metallurgy.

Copper is one of the rare metals that occur naturally in a pure metallic form. From about the 9th millennium BC, societies began to work copper by hammering it into desired forms. In this way, jewelry, tools and weapons were created. The oldest known copper ornament is dated to 8,700 BC. It was found in Mesopotamia, in the area of what is now northern Iraq.[159]

It was probably noted by the first workers of copper that heating it made it malleable and easier to work. The first liquefaction of copper represents a significant step in human development. Once liquified, copper could be poured into molds which allowed the fabrication of copper objects in complex shapes. However, copper melts at a temperature of 1,089° Celsius (1,985° Fahrenheit). How could such a temperature be reached?

Two technologies were required to bring copper to its melting point: charcoal and bellows. Charcoal is the product of wood that has been heated and partially burned in a low oxygen environment. Relative to wood, charcoal has a very low moisture content and produces high and steady quantities of heat. Bellows are devices that replicate the effect of blowing on a fire. Although the inventors of bellows would not have known it, bellows simply provide more oxygen for fuel to combust.

The Sumerians were the first people to have made wide use of copper, which they introduced to the Egyptians. The Egyptians used polished copper mirrors – the first mirrors – and also made use of

---

[159] **A Timeline of Copper Technologies** – Copper Development Association Inc – https://www.copper.org/ – Accessed: August 2020

copper-derived blue pigment for art and makeup. Copper tools such as axes for churning farmland increased in use and thereby increased agricultural yields. Copper tools also gave rise to carpentry.

The rolling of a circular object does not require brilliance. However, the invention of a wheel that can rotate on an axle does. Moreover, actually creating such a working system requires a mastery of the craft of carpentry and good tools. This is because where the rotating wheel makes contact with the stationary axle requires a near-perfect fit. According to David Anthony, a professor of anthropology who is an expert in the invention of the axled-wheel, "It was the carpentry that probably delayed the invention until 3500 B.C. or so, because it was only after about 4000 B.C. that cast copper chisels and gouges became common in the Near East."[160] Fire and the copper derived from fire had transformational effects on human development during this time. With wheeled carts it was no longer required for heavy loads to be borne on the backs of humans or domesticated animals, and copper was only the beginning of metallurgy.

It was bronze that completely transformed metallurgy and the course of human development. Relative to bronze, copper was a mere precursor.

The earliest evidence of bronze was discovered at Plocnik, Serbia, and is dated to circa 4,500 BC.[161] At its origins bronze use was marginal. However, from 3,000 BC to 1,200 BC in the area of the Eastern Mediterranean, Egypt, Mesopotamia and eastwards as far as present-day Afghanistan the use of bronze was so pronounced that that period of 1,800 years is referred to as that area's Bronze Age.[162]

Bronze is significantly harder than copper and no more difficult to make. Tin melts at 232° Celsius (449° Fahrenheit); it can be liquified on a stove top. Once added to copper in a weight ratio of roughly one unit of tin for every ten units of copper, bronze is produced: Bronze is an alloy made of copper and tin. A higher tin ratio produces a

---

[160] **Why it took so long to invent the wheel** – Natalie Wolchover – Scientific American – 6 March 2012

[161] **Tainted ores and the rise of tin bronzes in Eurasia, c. 6500 years ago** – Miljana Radivojevic, Thilo Rehren, Julka Kuzmanovic-Cvetkovi, Marija Jovanovic and J. Peter Northover – Antiquity – 87(2013)

[162] **1177 BC, The Year Civilization Collapsed** – Eric H. Cline – Princeton University Press – 2014

harder, but more brittle bronze alloy. The one-to-ten ratio became relatively standardized throughout the area where bronze was first widely used.

The principal difficulty in the production of bronze during the Bronze Age was the procurement of tin, which is much less abundant than copper. The growing demands for copper during the Bronze Age were easily met by mines on the island of Cyprus. The word copper is actually derived from the Greek word for that island, Kupros.[163] In contrast, tin was difficult to procure. Tin in the volumes required to support a thriving Bronze Age could only be supplied by the mines of the Badakhshan region in modern-day Afghanistan.[164]

The advantages derived from having bronze were so considerable that the nine independent civilizations that existed in the area were essentially compelled into trading with each other in order to acquire the copper and tin required to make bronze. By the end of the Bronze Age, these independent civilizations consisted of the Egyptian, Babylonian, Assyrian, Canaanite, Cypriot, Mitanni, Hittite, Minoan and Mycenaean civilizations.

The advantages of bronze combined with the benefits of international trade created a golden era of prosperity.

Bronze axes significantly increased agricultural yields. The use of ploughs became widespread for the first time due to their resistance to breaking when reinforced with bronze. Higher agricultural yields resulting from the use of bronze agricultural equipment created a surplus of food, which allowed for more urbanization and created the basis for the development of prosperity.

Bronze tools allowed for fine carpentry: Spoked-wheeled chariots were first produced during the Bronze Age. Professional armies appeared for the first time. They were of course kitted out with mass-produced weapons and armor made in bronze. The wealth created during this era allowed the Egyptians to construct the great pyramids

---

[163] **Cyprus-Island of Copper** – The Metropolitan Museum of Art – https://www.metmuseum.org – Accessed: August 2020
[164] **1177 BC, The Year Civilization Collapsed** – Eric H. Cline – Princeton University Press – 2014

while fostering the flourishment of writing and art throughout the region.

Tin underpinned the commerce, prosperity and social structures of the Bronze Age. Carol Bell, a research associate at the University of London's Institute of Archaeology, made the point that the strategic importance of tin in supporting the prosperity of Bronze Age civilizations was comparable to the strategic importance of oil today.[165]

Of the first nine Bronze Age civilizations, eight collapsed precipitously into complete ruin around 1,200 BC. Only the Egyptian civilization would survive to limp along in a greatly transformed and diminished capacity. Egypt would never again construct anything remotely close to the grandeur of the pyramids.[166] Effectively, over a period perhaps spanning no more than a century, the cultures, writing, economies and political structures of the Bronze Age civilizations ceased to exist.

The following factors contributed to the ruin of these great civilizations:[167]

i) earthquakes;

ii) invasions and rebellions;

iii) a collapse in trade and in particular a collapse in the supply of tin; and

iv) climate change and a prolonged period of resulting famine.

We will now look at the identified causes of the collapse of the Bronze Age civilizations to draw out implications for our own societies:

i) Archaeologists believe that many of the ruins of the period between 1225 BC and 1175 BC have the signature of

---

[165] **The merchants of Ugarit: oligarchs of the Late Bronze Age trade in metals?** – Carol Bell – in Eastern Mediterranean Metallurgy and Metalwork in the Second Millennium BC – Oxbow Books – May 2012

[166] **1177 BC, The Year Civilization Collapsed** – Eric H. Cline – Princeton University Press – 2014

[167] **1177 BC, The Year Civilization Collapsed** – Eric H. Cline – Princeton University Press – 2014

earthquake damage. The implication is that successful societies must be strong enough to withstand unexpected challenges.

ii) The civilizations of the Bronze Age were under attack by the so-called Sea Peoples of the Mediterranean Sea. Recent thinking suggests that the Sea Peoples were refugees or victims of the collapse of the Bronze Age rather than the instigators of that collapse.[168] The experience of the Bronze Age civilizations suggests that prosperity creates stability and that extreme poverty creates instability. The tragedies that are currently unfolding on the Mediterranean Sea are reminders that extreme poverty, hardship and insecurity remain far too pervasive on our Earth. At least 22,400 people are estimated to have lost their lives trying to enter Europe since 2000, mostly in the Mediterranean Sea.[169] Consideration to the poorest amongst us will be particularly relevant when we assess the economic implications of replacing fire with alternatives. This will be addressed in the section "Costs of Fire vs. Alternatives to Fire".

iii) The collapse in the trade of tin took the bronze out of the Bronze Age. If archaeologist and historian Carol Bell is correct that the strategic importance of tin in the Bronze Age was comparable to that of oil today,[170] we should be very thoughtful about the implications of prematurely eliminating our use of oil.

iv) Rhys Carpenter was an American archaeologist, art historian and man of letters. In 1966, he argued that drought and a resultant famine were important root causes of the calamity that occurred at the end of the Bronze Age.[171] Science recently corroborated his argument: From 2012, four

---

[168] **1177 BC, The Year Civilization Collapsed** – Eric H. Cline – Princeton University Press – 2014
[169] **Fatal Journeys, Tracking Lives Lost during Migration** – Editors: Tara Brian and Frank Laczko – International Organization for Migration – 2014
[170] **The merchants of Ugarit: oligarchs of the Late Bronze Age trade in metals?** – Carol Bell – in Eastern Mediterranean Metallurgy and Metalwork in the Second Millennium – May 2012
[171] **Discontinuity in Greek Civilization** – Rhys Carpenter – Cambridge University Press – 1966

different teams of archaeological researchers have provided paleo-climactic data indicating that there was a prolonged period of drought that coincided with the period of the collapse of civilizations at the end of the Bronze Age.[172] The implications are twofold: Firstly, in contrast to widely held perceptions that science is at the forefront of our knowledge, a man of letters understood the climate history of our Earth almost a half century before science was able to catch up. Secondly, the period of reduced precipitation at the end of the Bronze Age was caused by a cooling event – the opposite of the current warming event. The Earth is currently becoming wetter and more humid, not drier. This will be discussed in detail in the section "From Fire: Carbon Dioxide Fertilization".

In contrast to the prosperity and magnificence of the Bronze Age, the Iron Age that followed was for several centuries a period of darkness characterized by hardship, illiteracy and subsistence farming.

From Fire: the Iron Age

As with the Bronze Age, fire was essential to the Iron Age. The production of iron and steel are not only dependent on the heat produced by fire, but also on the chemicals (carbon and carbon monoxide) that are produced by fire. This will be further elaborated in this section.

The civilizations that emerged at the end of the Iron Age were built-up from rudimentary conditions closer to those of the Stone Age than the magnificence of the Bronze Age. The region's Iron Age began

---

[172] **Climate and the Late Bronze Collapse: New Evidence from the Southern Levant** – Dafna Langgut, Israel Finkelstein and Thomas Litt – Journal of the Institute of Archaeology of Tel Aviv University – 17 February 2014;
Late Bronze Age climate change and the destruction of the Mycenaean Palace of Nestor at Pylos – Martin Finné, Karin Holmgren, Chuan-Chou Shen, Hsun-Ming Hu, Meighan Boyd, Sharon Stocker and John P. Hart – PLoS One – 27 December 2017;
The influence of climatic change on the Late Bronze Age Collapse and the Greek Dark Ages – Brandon Lee Drake – Journal of Archaeological Science – June 2012; and
Environmental Roots of the Late Bronze Age Crisis – David Kaniewski, Elise Van Campo, Joël Guiot, Sabine Le Burel, Thierry Otto and Cecile Baeteman – 14 August 2013

with the collapse of the Bronze Age and ended around 500 BC, which roughly coincides with the foundation of the Roman Republic and the beginning of recorded history.

Iron is in most respects better than bronze. This was particularly the case for the applications of metallurgy in antiquity. Iron can be made stronger and harder than bronze. A sword made of iron will be less likely to break and will keep a sharp edge for longer than a sword made of bronze. Moreover, the abundance of iron made its supply more secure than that of tin. The wide abundance of iron also meant that it had potential to democratize metallurgy and to provide benefit to everyone, not just the rulers and elite who could control the supplies of tin. Why then did the end of the Bronze Age and the onset of the subsequent Iron Age not quickly propel humanity into an age of unprecedented prosperity?

The answer is that the technologies to exploit iron had not yet been sufficiently developed.

This consideration brings into focus the implications of prematurely replacing the human use of fire, namely, fire from coal, oil and natural gas, with alternatives to fire before those alternatives are sufficiently technologically advanced.

Let us take a closer look at the character of iron and the development of its use over the Iron Age.

Iron has a melting point of 1,538° Celsius (2,800° Fahrenheit). For most of the Iron Age, furnaces were insufficiently hot to melt iron. The first evidence of casting liquid iron is dated 645 BC and located in China.[173]

In contrast with copper, iron does not exist naturally in a metallic form. Iron must be derived through a process called smelting. Smelting transforms minerals that have a high iron content into something close to pure iron. Smelting involves removing the components of iron ore that are non-metallic through chemical reactions that occur at high temperatures. Fire, not just heat, is required to smelt iron ore because fire provides the carbon and carbon monoxide that are required to reduce the iron ore into iron. At

---

[173] **A History of Cast Iron, ASM Handbook, Cast Iron Science and Technology** – Doru M. Stefanescu – ASM International – 2017

its origins smelting also involved the repeated hammering of red-hot ore to remove unwanted impurities and to consolidate the ore. The result of this process is a relatively pure form of iron called wrought iron.

Wrought iron is not particularly advantageous relative to bronze; it rusts easily and is not appreciably harder. To fully exploit the potential of iron requires it to be carburized, which is the absorption of carbon into iron through heat-related processes. The product of carburation is steel. Steel is not only the product of heat from fire, but of the carbon that is produced from fire.

Carbon can be introduced into iron during the smelting process or later. Once iron has been carburized into steel, it is receptive to subsequent heating, cooling and hammering processes that will permanently alter its character.[174] Although the elements of steel-craft are basic, the sequencing and degree to which each process is applied requires a high level of skill to achieve the desired outcome. Steel-craft is unforgiving of errors because its processes are irreversible.

The urban centers of Etruria would become the most prolific producers of iron and steel in the area of the Mediterranean in the millennia that followed the collapse of the Bronze Age civilizations. Etruria was located on the central-west coast of what is now Italy. The peoples of ancient Etruria are called Etruscans. The area is distinguished in the Mediterranean region for its large-scale iron ore resources. Iron artefacts in the area dated from between the collapse of the Bronze Age civilizations and circa 800 BC are few and isolated. However, from 800 BC onwards, iron use began increasing significantly as meaningful amounts of iron started being produced.[175]

As the use of iron ploughs for farming became more prevalent, agricultural yields increased.[176] The resulting surplus food allowed

---

[174] **A History of Cast Iron, ASM Handbook, Cast Iron Science and Technology** – Doru M. Stefanescu – ASM International – 2017
[175] **Metallurgy and the Development of Etruscan Civilisation** – Pieter William Mommersteeg – Dissertation, Ancient History, University College London – 2011– Accessed via UCL discovery (August 2020)
[176] **The Urbanisation of Rome and Latium Vetus, From the Bronze Age to the Archaic Era** – Francesca Fulminante – Cambridge University Press – 2014

small villages with thatch-roofed huts to emerge throughout the area.[177] Essentially, the processes that led to the prosperity of the Bronze Age were repeated.

Seaborne trade on the Mediterranean recommenced in a meaningful way, which led to the exchange of technologies and ideas. Notably, Greek migrants spread throughout the Mediterranean bringing with them advanced metallurgy and philosophy. From the hardship of the dark ages, prosperity gradually emerged throughout the region. That prosperity was not only for the elite rulers or the wealthiest merchants. Throughout the area progress raised living standards well above subsistence levels for millions of people. People lived longer, ate better, had more comfortable homes and enjoyed more numerous, more varied and higher quality goods.[178]

In the year 509 BC, the citizens of a town, inspired by Greek political philosophy, overthrew their Etruscan king and declared their town to be "res publica", an entity belonging to the people – a republic. That town was Rome – at last civilization was finally and meaningfully on the up.

It is said that those who do not learn from history are doomed to repeat it. Bronze Age and Iron Age history tells us that materials are important for establishing and maintaining prosperity. If archaeologist and historian Carol Bell is correct that the strategic importance of tin for Bronze Age civilizations was similar to that of oil today, we had better make sure that the replacement of oil works not only in theory, but in practice as well.[179] Iron, although theoretically superior to bronze, required centuries of technological developments before serving as a satisfactory replacement to it. The centuries during which iron was unable to satisfactorily replace bronze were dark ages that followed a period of tremendous prosperity.

---

[177] **The Archaeology of Early Rome and Latium** – R. Ross Holloway – London Routledge – 1996
[178] **The Cambridge Economic History of the Greco-Roman World** – Walter Scheidel, Ian Morris and Richard Saller – Cambridge University Press – 2007
[179] **The merchants of Ugarit: oligarchs of the Late Bronze Age trade in metals?** – Carol Bell – in Eastern Mediterranean Metallurgy and Metalwork in the Second Millennium – May 2012

Fire underlies many of the materials that we use to create prosperity, inclusive of metals.

Turning our focus of attention to the present, in 2018, global steel production amounted to 1,809 million metric tons.[180] For every ton of copper required to support our global economy, we require about 75 tons of steel.[181]

Today coal, not woodfuel, is used to smelt iron and to create steel. To produce one ton of steel from iron ore requires 0.77 tons of coal on average.[182]

Currently, about 75% of steel is produced beginning-to-end from fire-based processes. The remaining 25% of steel production uses electricity for some of the production processes. However, smelting iron ore into iron requires fire as does the production of steel.[183] In practical terms, steel is derived from fire, namely, fire from burning coal.

The European Union, and in particular the European Commission, has been a driving force in favor of the suppression of the human use of fire in favor of alternatives.[184] The European Commission evolved from the High Authority of the European Coal and Steel Community.[185] For reference, no representatives of the European Commission are directly elected by the citizens of Europe.[186] The politics of energy in Europe will be discussed further in the section "From Fire: Politics". The history of the European Union's origins is relevant as it relates to both steel and the human use of fire, namely, fire from burning coal.

---

[180] **World Steel in Figures** – World Steel Association – https://www.worldsteel.org/ – 2019
[181] **World Refined Copper Production and Usage Trends** – International Copper Study Group – http://www.icsg.org/ – Accessed: August 2020
[182] **Metallurgical Coal** – BHP – https://www.bhp.com – Accessed: February 2021
[183] **Energy use in the steel industry** – World Steel Association – April 2019
[184] **A European Green Deal** – European Commission – https://ec.europa.eu – Accessed: March 2021
[185] **The European Commission, Fact Sheets on the European Union** – European Parliament – https://www.europarl.europa.eu – Accessed: March 2021
[186] **About the European Commission** – The European Commission – https://ec.europa.eu – Accessed: March 2021

After the Second World War, steel and energy were recognized as critical enablers to create both prosperity and security. For this reason, in 1951, a treaty was signed by six European countries that created the European Coal and Steel Community. The primary purpose of that treaty and the mission of the institutions created by that treaty was to ensure that signatory countries would be well supplied with coal and steel at the lowest prices possible (article 3). The foundational principles of that treaty were the freedom of trade within the community, restrictions on governmental interventions in markets and the promotion of equitable trade with third party countries.[187] That treaty played an important role in creating what the French, who were one of the six signatories, refer to as "Les Trentes Glorieuses", referring to the thirty years of prosperity, rising living standards and optimism experienced in Europe from 1945 to 1975. The institutions created by that treaty have evolved to become those of the European Union.

Focusing on the present, the European Union has imposed directives and regulations on its member countries that suppress the use of coal due to the carbon dioxide it emits. As a result, coal consumption by the member countries of the European Union fell by 14% from 2000 to 2018.[188] The corollary of this change is that the European Union is no longer able to competitively produce the steel that it requires. Table 2 shows that the European Union has become the world's largest importer of steel. According to the European Commission itself, the steel industry in the European Union is in decline because i) up to 40% of the manufacturing costs of steel production are energy costs and ii) "European industry is faced with higher energy prices than most of its international competitors."[189] Paradoxically, rather than eliminating coal consumption and carbon dioxide emissions, the policies of the European Union have simply transferred steel production and coal consumption to steel-exporting countries, namely, China, Japan and Russia.

The treaty establishing the European Coal and Steel Community was valid for 50 years and expired in 2002. It is a historical irony that the

---

[187] **Treaty Establishing the European Coal and Steel Community** – 1952
[188] **Energy balance sheets, 2020 Edition** – Eurostat – 2020
[189] **The EU steel industry** – European Commission – https://ec.europa.eu – Accessed: August 2020

institutions created for the explicit purpose of guaranteeing the supply of low-cost coal and steel are now a global driving force suppressing the use of coal and the associated production of low-cost steel.

Top importers & exporters of steel 2019 *Table 2*

Steel Importers		Steel Exporters	
Country	Mt	Country	Mt
EU 28	45	China	69
US	32	Japan	36
Thailand	16	Russia	33

Source: World Steel Association[190]

In this section, we have looked at the basic materials that have formed the foundations of our civilization.

With materials such as steel, humanity has lifted itself into the skies. Today, the Burj Khalifa in Dubai is the tallest building in the world. Gleaming with glass and steel, it stands 828 meters (2,716 feet) tall. It is in every respect as magnificent as the great constructions of the ancient world. However, even with all the steel and glass in the world, the splendor of the Burj Khalifa would not have been possible without one thing: the elevator. In the next section, we will look at how fire transformed our world, only quite recently, by making things move up and down, rotate and travel.

---

[190] **World Steel in Figures** – World Steel Association – https://www.worldsteel.org/ – 2019

3. From Fire: Energy

In the period that followed the Iron Age, the Romans were ascendant. The iconic Roman sword, the gladius, was made in steel. It was perfectly designed for Roman warfare. It was versatile and, unlike many sword designs, it could be used to make forward stabbing thrusts overtop of a Roman shield.

The Romans and their gladii ultimately conquered the entirety of the coastal regions of the Mediterranean Sea and much of Europe, Northern Africa and Asia Minor.

The republican principle that citizens are the source of governmental authority was never fully applied in Rome, albeit powers were divided amongst the elite to ensure no person would retain complete power. Ultimately, the founding republican principles of Rome were betrayed and, from 27 BC, Roman power was concentrated in the hands of the ruling emperor. Corrupt and inept leadership combined with the onslaught of military offences from barbarian tribes, including the Visigoths, Vandals, Angles, Saxons, Franks, Ostrogoths and Lombards led to the collapse of the Romans. In the year 476 AD the last gladius fell. The Roman Empire had been defeated.

After another long period of dark ages, the most recent pulse of human progress has largely been a struggle to apply and broaden the republican principles of early Rome. The most prosperous and powerful countries today have democratized republican ideals. Premised on the freedom of all individuals, their equality and the republican principal that the people collectively give authority for governments to have and to use power, modern societies have created an unprecedented period of progress, prosperity and peace.

In recent centuries individual freedoms in respect of economic activity have formed the basis of free market principles.

The political and economic systems that emerged through the struggle for freedom created the stimulus for one of the most important human developments since the Agricultural Revolution: the Industrial Revolution.

Fire was again the essential enabler of this phase of humanity's progress.

During the Industrial Revolution, fire was not just used to cook, produce materials and for lighting, but, for the first time, to make things move.

The history of fire since the beginning of the Industrial Revolution has been in large part the history of the evolution of fuels that have been burned to produce energy – from wood to hydrogen. This section will cover the principal fuels used to provide our energy from the beginning of the Industrial Revolution to the present. It will be divided into sub-sections by fuel type, starting with wood. It will end by covering forms of energy that do not involve fire.

Fire from Wood, Energy for the Billions

The first humans burned wood for heat, light and cooking. Contrary to widely held perceptions, fire from wood remains one of the most important forms of energy globally.

Citing directly from the 2018 United Nations' State of the World's Forests report, "Woodfuel – defined as including both fuelwood and charcoal – is used by approximately 2.4 billion people worldwide for cooking meals, sterilizing drinking water and heating homes … Overall, forests supply about 40% of global renewable energy in the form of woodfuel – as much as solar, hydroelectric and wind power combined."[191]

Since the Industrial Revolution, in industrialized countries fire from burning coal has replaced fire from burning wood. Coal and the beginning of the Industrial Revolution are the subjects we will look at next. The first countries to have industrialized were European.

Contrary to widely held perceptions, both industrialization and the industrialized use of fire from burning coal, oil and natural gas have been the drivers that have allowed European forests to prosper. There are multiple reasons for this; one key reason is that replacing fire

---

[191] **The State of the World's Forests 2018** – Food and Agricultural Organization of the United Nations – 2018

from burning wood with fire from burning coal reduced pressures on forests for woodfuel.

The effects of industrialization on forests are best evidenced by looking at the history of the largest European country by surface area, France. France was a feudal agricultural society until 1830. In that year, France appointed a new king, Louis Philippe the 1st, who sought to replicate the industrial success of the United Kingdom. The Industrial Revolution in France started in the year 1830. From 1830 to 2018, the forested surface area of France actually *increased* by 69%.[192] This trend is ongoing: Between 1950 and 2010, the forested area of the European Union (including the United Kingdom and Switzerland) grew by 25.4%.[193]

Similar comparisons are not possible in countries like the United States where pre-industrial agricultural-based societies never existed. For reference, the United States was inhabited by nomadic hunter-gatherers before being settled by industrialized Europeans. However, the implications of Europe's industrialization on wildlife habitat are universal.

In contrast to widely held perceptions, industrialization and the fire that powers industry are blessings that have protected wildlife habitat and wildlife – even as the global population has grown to 7.7 billion. Although this is not widely recognized, this is *by far* the most important consideration in relation to how stopping the human use of fire will affect wildlife.

As an important digression, science has only recently acknowledged that changes to landscapes, particularly vegetative landscapes such as forests, affect the climate. This will be discussed in the section "From Fire: Carbon Dioxide Fertilization". The important point to retain is that changes to the Earth's vegetative landscapes can be driven by a number of factors, even by industrialization.

---

[192] **La Forêt en France Métropolitaine** – Institut National de l'Information Géographique et Forestière – http://education.ign.fr/dossiers/foret-france-metropolitaine – Accessed: August 2020
[193] **A high-resolution and harmonized model approach for reconstructing and analysing historic land changes in Europe** – Richard Fuchs, Martin Herold, Peter Verburg, J.G.P.W Clevers – Biogeosciences – March 2013

Before leaving the subject of fire from woodfuel, it is critical to appreciate that the real energy challenge for the decades and centuries ahead relates to the provision of better forms of energy to the 2.4 billion people who currently cannot afford energy other than the woodfuel provided by forests.

Fire from Coal, the Embattled King

Fire from burning coal powered the beginning of the Industrial Revolution and, today, provides 26.9% of global energy.[194]

To understand coal requires an understanding of its origins, which brings us back in time – way back. The bulk of the coal driving the Industrial Revolution and contributing to global warming is derived from peat that was deposited during the Carboniferous Period (359–299 million years ago).[195]

During the beginning of the Carboniferous Period, the climate globally was generally hotter, more humid and more tropical than our current climate.[196] Also at the beginning of the Carboniferous Period, carbon dioxide made up more than 0.1% of the atmosphere.[197]

The formation of coal reduced the amount of carbon dioxide in the atmosphere over the course of the Carboniferous Period.[198] This is because during the Carboniferous Period large amounts of peat accumulated on Earth, which had the effect of capturing carbon and burying it – much like the formation of peat today as discussed in the section "Peatfires and Peat". The peat that accumulated during the

---

[194] **World Energy Balances 2020 (Data for 2018)** – International Energy Agency – July 2020
[195] **Formation of most of our coal brought Earth close to global glaciation** – Georg Feulner – Proceedings of the National Academy of Sciences of the United States of America – September 2017
[196] **The Carboniferous Period** – University of California Museum of Palaeontology, Berkley – https://ucmp.berkeley.edu/ – Accessed: August 2020
[197] **CO2 as a primary driver of Phanerozoic climate** – Dana L. Royer, Robert A. Berner, Isabel P. Montañez, Neil J. Tabor, David J. Beerling – GSA Today – March 2004
[198] **CO2 as a primary driver of Phanerozoic climate** – Dana L. Royer, Robert A. Berner, Isabel P. Montañez, Neil J. Tabor, David J. Beerling – GSA Today – March 2004

Carboniferous Period was transformed into coal through natural geological processes.[199]

By the end of the Carboniferous Period, the amount of carbon dioxide in the atmosphere had declined by ten times and made up as little as 0.01% of the atmosphere.[200] As a result, the Earth experienced its coldest glacial period of the last 300 million years at the end of the Carboniferous Period. Researchers believe that during this time the Earth only narrowly escaped from a "snowball" state with the equators entirely covered in ice sheets.[201]

As an important digression, in 2016, research, based on the first-ever high-definition geological data, indicated that changes to the amount of carbon dioxide in the atmosphere during the Carboniferous Period affected the Earth's vegetation, which in turn had feedback effects on the climate during that period[202] – climate science is changing and fast.

Before the Industrial Revolution, carbon dioxide made up 0.028% of the atmosphere.[203] Fire from burning coal has contributed materially to rising levels of carbon dioxide. As a result, currently, carbon dioxide makes up 0.0417% of the atmosphere.[204]

Fire from burning coal is reversing the effects of the formation of coal during the Carboniferous Period by releasing carbon dioxide back into the atmosphere. As a result, our Earth is becoming less like the frozen Earth that existed at the end of the Carboniferous Period

---

[199] **Coal** – National Geographic – https://www.nationalgeographic.org – Accessed: February 2021
[200] **Formation of most of our coal brought Earth close to global glaciation** – Georg Feulner – Proceedings of the National Academy of Sciences of the United States of America – September 2017
[201] **Formation of most of our coal brought Earth close to global glaciation** – Georg Feulner – Proceedings of the National Academy of Sciences of the United States of America – September 2017
[202] **Climate, pCO2 and terrestrial carbon cycle linkages during late Palaeozoic glacial–interglacial cycles** – Isabel P. Montañez, Jennifer C. McElwain, Christopher J. Poulsen, Joseph D. White, William A. DiMichele, Jonathan P. Wilson, Galen Griggs and Michael T. Hren – Nature Geoscience – 24 October 2016
[203] **The Atmosphere: Getting a Handle on Carbon Dioxide** – Alan Buis – NASA – https://climate.nasa.gov – Accessed: March 2021
[204] **Monthly Average Mauna Loa CO$_2$** – US National Oceanographic and Atmospheric Administration – https://www.esrl.noaa.gov – Accessed: March 2021

and more like the tropical Earth that existed at the beginning of the Carboniferous Period.

Refocusing our attention on the history of coal, before 1698, fire was not used to make things move, but rather for cooking, warmth, light and to make materials. However, in that year, Thomas Savery of England invented a steam-driven water pump that was heated by the fire of burning coal. In 1765, James Watt, a Scottish inventor, significantly improved the steam engine design. Importantly, he developed a system of gears that was able to transform reciprocating (up-down) motion into rotational motion – the Industrial Revolution, fueled by the fire of burning coal, was full steam ahead.

Steam engines, driven by the heat of fire from burning coal, powered the expansion of industry and railways. Steam engines also allowed ships to replace their sails with propellers.

Coal was king. It provided the propulsive force of the Industrial Revolution. Clearly, today, coal is not the king. What happened?

The symbolic moment that dethroned coal occurred when Winston Churchill, appointed First Lord of the Admiralty prior to the First World War, decided that the strength of the United Kingdom's Royal Navy would increase if its ships were fueled by oil rather than coal. Although switching from coal to oil meant that that navy would be dependent entirely on distant, foreign oil, the change was considered to be worth it because oil was simply that much better.[205]

Focusing on the present, selected coal consumption data is provided in Table 3.

---

[205] **The Prize** – Daniel Yergin – Simon & Schuster – 1991

Coal consumption for energy (Mt) — Table 3

	1990		2013		2019	
China	709	23%	2,922	52%	2,866	53%
India	132	4%	488	9%	585	11%
USA	666	21%	614	11%	397	7%
EU	625	20%	399	7%	250	5%
Other	986	32%	1,168	21%	1,309	24%
Global	**3,118**	**100%**	**5,591**	**100%**	**5,407**	**100%**

Source: International Energy Agency[206]

In 2019, in addition to the coal used to for energy, 1,080 million metric tons of coal were used to produce steel.[207]

Coal-fired power plants emit nitrous oxides, sulfur dioxide, mercury, fine particle pollution and other pollutants.[208] Measures to reduce pollutants from coal-fired power plants have proven effective in the United States.[209] Pollution will be discussed further in the section "From Fire: Pollution".

Fire from Oil, to Win

Oil is currently the king of energy; fire from burning oil provides 31.5% of global energy.[210]

Oil is formed similarly to coal, with some important differences. A considerable amount of the oil we use was deposited as organic plant matter within mud at the bottom of ancient seas and oceans. This mud turns into rock, specifically shale, when it is buried to a sufficient depth by the accumulation of overlying sediments. If organic-rich shale is buried to within certain depths for a long enough period of time, the organic plant matter within shale transforms into oil. The

---

[206] **Coal Information: Overview** – International Energy Agency – Statistics Report July 2020
[207] **Coal 2020** – International Energy Agency – December 2020 – https://www.iea.org – Accessed: February 2021
[208] **Cleaner Power Plants** – United States Environmental Protection Agency – https://www.epa.gov – Accessed: February 2021
[209] **Reducing Air Pollution from Power Plants: The Success of the Mercury and Air Toxics Standards** – American Lung Association – 14 March 2019 – https://www.lung.org – Accessed: February 2021
[210] **World Energy Balances 2020 (Data for 2018)** – International Energy Agency – July 2020

genesis of oil is an expansive process. It creates pressure as the volume of oil in the shale is increased. Ultimately, oil gets expulsed from shale as a liquid, perhaps several kilometers (miles) under the surface of the Earth. In this respect oil is nothing like coal, which does not move and remains a solid rock. Once expulsed from shale, oil tends to rise upwards. This is because within sedimentary rock, where oil originates, water typically fills the spaces between rock grains. Oil is lighter than water and its buoyancy pushes it to the surface of the Earth. Occasionally, oil's upward rise is blocked by a layer of impermeable rock. A highly porous sedimentary layer of sandstone shaped like a dome over which impermeable rock is draped would be perfect for catching oil trying to rise to the surface of the Earth. Such a geological setting would be representative of many of the world's largest oil discoveries.[211]

Oil wells, which consist mainly of steel tubes, produce oil that has been trapped in the subsurface. Oil wells are drilled from the surface and they penetrate into the porous rocks where oil is trapped. Oil flows from porous rocks into the steel tubing and from there up to the surface.

Oil is also called petroleum, which is derived from the Latin words "petra" for rock and "oleum" for oil. Gasoline, jet fuel and diesel are oil-derived fuels. Those fuels are burned in internal combustion engines. Heat from fire is the driving force of all internal combustion engines.

In addition to producing energy, many synthetic materials are derived from oil, inclusive of synthetic fibers. Professional athletes are typically seen head-to-toe in clothes that are made in materials derived from oil. Athletes choose synthetic materials because they are more comfortable, lighter, more breathable, quicker to dry and generally higher performing materials than cotton, the most abundant of farm-produced fibers.

---

[211] **Elements of Petroleum Geology** – Richard C. Selley and Stephen A. Sonnenberg – Elsevier – 2014

Synthetic oil-derived fibers currently have a market share of 62% of the fiber market.[212] The use of oil-derived synthetic materials reduces the land required to produce farmed materials. For example, globally, a surface area of 324 thousand square kilometers (125 thousand square miles) is used to grow cotton.[213] That equates to 47% of the surface area of the state of Texas, USA. The land being used to grow cotton reduces the availability of land for wildlife. By using synthetic fibers, derived from oil, more land is made available for wildlife.

Oil-derived products are ubiquitous, from surfboards and the components of mobile phones to products that are less obviously linked to oil, such as paints and water pipes. Plastics derived from oil are inexpensive, light and durable. Sadly, because of this, oil-derived plastics are contributing to pollution in our oceans and harming ocean wildlife.

One of the most comprehensive studies on plastic ocean waste determined that China, Indonesia and the Philippines are the worst plastic polluters into our oceans.[214] The European Union countries combined rank as the 18th worst polluter and the United States ranks as the 21st worst polluter. The scale of China's plastic pollution into our oceans is 31.5 times greater than that of the United States. For reference, the United States is 6.3 times more prosperous than China based on GDP per capita.[215] Research indicates that prosperity is associated with lower levels of plastic pollution. The relationship between prosperity and low pollution levels will be seen again in the section "From Fire: Pollution". Critically, the use of fire has been the driving force that has increased human prosperity from our origins. Prosperous countries pollute less.

---

[212] **Preferred Fiber & Materials Market Report 2019** – Sophia Opperskalski, SuetYin Siew, Evonne Tan, Liesl Truscott – Textile Exchange – https://textileexchange.org/ – Accessed: August 2020
[213] **Measuring Sustainability in Cotton Farming Systems** – Food and Agricultural Organization of the United Nations – 2015
[214] **Plastic waste inputs from land into the ocean** – Jenna R. Jambeck, Roland Geyer, Chris Wilcox, Theodore R. Siegler, Miriam Perryman1, Anthony Andrady, Ramani Narayan, Kara Lavender Law – Science – 13 February 2015
[215] **GDP per Capita, 2019** – World Bank – https://data.worldbank.org – Accessed: November 2020

Although oil is used to make materials, its primary use is to make things move. Cars, trains, planes and boats are all powered by the fire of burning oil.

In 1885, Carl Benz, a German engineer and entrepreneur, was the first person to develop and market a working automobile powered by the internal combustion engine – driven by the heat of fire produced from burning oil. In the United States, the Model T automobile developed by Henry Ford had the objective of "putting the world on wheels" by making cars affordable to the general public. It was first produced in 1908.

The affordability of automobiles and the oil used to fuel them is their quintessential characteristic. Automobiles are for the benefit of everyone. Automobiles provided normal people with extraordinary personal freedoms.

The importance of fire from oil as a form of energy to make things move is best appreciated from the perspective of the role it played in the most pivotal challenge of the last century: the Second World War.

According to historians, the Second World War was defined by the struggle for a single resource – oil.[216] Importantly, nations do not fight wars over oil simply for its own sake, but rather to accomplish the tasks that require oil.[217]

Hitler's "blitzkrieg" or lightning war strategy had to be fierce but short before Germany ran out of oil. Germany invaded Russia to gain access to eastern oilfields. However, that strategy failed and resulted in the further depletion of its oil stocks. As a result, according to Captain Mawn of the United States Navy, "The German Army had the greatest use of equestrian transport of any military conflict in history. Typically, in the last two years of the war, German Luftwaffe airplanes were pulled to runways from aviation hangers and parking locations by horses, cows and oxen."[218] During the Second World War Germany ran out of oil.

---

[216] **How Oil Defeated the Nazis** – Gregory Brew – https://oilprice.com – 5 June 2019 – Accessed: August 2020
[217] **Oil & The Great Powers, Britain & Germany, 1914-1945** – Anand Toprani – Oxford University Press – 2019
[218] **Oil and War** – Captain Paul E. Mawn – https://defense.info/ – 24 October 2018 – Accessed: August 2020

In contrast, the victorious Allies were well supplied with American oil and the success of their efforts was linked to the logistics of getting that oil to the front lines. General Patton, who drove the United States Third Army across France to Germany after the Allied Invasion of Normandy in June 1944, told President Eisenhower, "My men can eat their belts, but my tanks have gotta have gas."[219] Frustrated by insufficient oil supplies as he advanced his army towards Germany, he wrote, "If I could only steal some gas, I could win this war."[220]

Growing up, during the winters, I would make money by shoveling snow off my neighbors' sidewalks. One of my long-time employers was Charles (Chuck) Mawer, who had been a Commander in the Royal Canadian Navy during the Second World War. He saw battle in the English Channel and escorted troop convoys to France on D-Day. He was awarded the Distinguished Service Cross for his actions. The nearest ocean to where I grew up was about 1,000 kilometers (620 miles) away by car, so at a Christmas party I asked Charles Mawer why he had joined the navy. He explained that the Allied navies were protecting the flow of war supplies from the United States and Canada across the Atlantic Ocean to Europe. When he explained the vital importance of the merchant marine which was shipping those supplies, I suggested that it seemed as though the Navy and the merchant marine won the war. He disagreed, and explained that it was true that the war would not have been won without the Navy and the merchant marine, "but we *all* won the war."

Captain Mawn makes the point about as clearly as possible: "Without American oil, the Second World War would not have been won by the Allies."[221] In the logic of Charles Mawer, the entirety of the service men and woman and the great many people who supported the war effort *all* won the war, but large-scale oil supplies were a necessary condition for that victory to be achieved.

Just as Winston Churchill switched the United Kingdom's Royal Navy from coal to oil, we can expect the leading armies and navies of the world to be extremely progressive when it comes to adopting

---

[219] **The Prize** – Daniel Yergin – Simon & Schuster – 1991
[220] **The Prize** – Daniel Yergin – Simon & Schuster – 1991
[221] **Oil and War** – Captain Paul E. Mawn – https://defense.info/ – 24 October 2018 – Accessed: August 2020

advantageous forms of energy. For purposes where another form of energy is superior to oil, oil has already been replaced. For example, the United States Navy launched the first nuclear powered submarine, the USS Nautilus, in 1954, four years before the construction of the first American civilian nuclear power station – we will know that a better fuel than oil has arrived when we see the leading armies and navies of the world aggressively adopting it.

Compared to other forms of energy, oil supports an extremely high tax burden, which materially improves the ability of governments to fund public services (Table 4). The tax burden supported by oil is consistent with the assessment that oil's strategic significance today in supporting our prosperity is comparable to that of tin during the Bronze Age.[222]

Gasoline tax duties	Table 4
	Fuel Duty Tax as Percent of Retail Price
USA	19.9%
UK	65.5%
Canada	32.0%
Australia	36.2%
France	63.9%
Germany	63.6%
Average of all OECD countries	**55.1%**

Source: OECD[223]

Importantly, oil can be produced with a negligible amount of land. For example, northern Alaska has produced, cumulatively, more oil through a single pipeline than all the oil produced by all countries combined during the last year of the Second World War.[224] That pipeline is designed to allow animals to graze underneath it (Figure

---

[222] **The merchants of Ugarit: oligarchs of the Late Bronze Age trade in metals?** – Carol Bell – in Eastern Mediterranean Metallurgy and Metalwork in the Second Millennium BC – Oxbow Books – May 2012

[223] **Taxation of premium unleaded (94-96 RON) gasoline (per litre), 2017** – OECD Tax Data Base – www.oecd.org – Accessed: September 2020

[224] **Oil & The Great Powers, Britain & Germany, 1914-1945** – Anand Toprani – Oxford University Press – 2019; and
Alaska Field Production of Crude Oil – US Energy Information Administration – https://www.eia.gov – Accessed: March 2021

11). Alaska's surface area is 4.8 times greater than that of Germany – one pipeline represents an insignificant amount of land in that context. Modern oil drilling operations in Alaska recover underground resources that have an area that is 6,500 times greater than the actual amount of land required on the surface (Figure 11).[225] In practical terms, for moose, grizzly bear and fish, humans have no presence at all in the oil producing regions of Alaska. Alaskan oil production not only protects wildlife habitat in Alaska, but globally, because Alaskan energy production averts the need to use extensive amounts of land elsewhere to meet our energy needs.

Minimal land use – oil operations in Alaska   *Figure 11*

12 Acre Gravel Pad

~ 125 sq. miles

Image Credit: ConocoPhillips (adapted; left) and US Bureau of Land Management (right)

Although the world is not at war, the conclusions relating to the effectiveness of fire from oil during the Second World War to achieve social and economic goals remain pertinent.

Fire from Natural Gas, Blue Flames

Fire from burning natural gas provides 22.8% of global energy.[226] Although not widely appreciated, natural gas is gaining global market share faster than any other form of energy.

---

[225] **ConocoPhillips Alaska: Investing in Alaska in Changing Times, Investor Presentation** – Joe Marushack – ConocoPhillips – 12 January 2017
[226] **World Energy Balances 2020 (Data for 2018)** – International Energy Agency – July 2020

If the natural processes that transform organic plant matter into oil occur deep enough in the Earth, the increased temperatures at which the processes occur start to create a combination of oil and natural gas. If these processes occur at extreme depths, no oil is produced and only natural gas is produced.[227]

Natural gas, like oil, is obtained from wells. Natural gas is transported from where it is produced to where it is consumed via underground pipelines.

The principal difference between natural gas and oil is just that, at normal temperatures natural gas is in a gaseous form. The volume required to store natural gas is 868 times greater than the volume required to store an equivalent amount of gasoline.[228]

Natural gas is unique amongst natural fuels in that it burns without producing anything other than carbon dioxide and water.[229] As a result of its perfect combustion, natural gas has a blue flame.

Compared to burning coal, burning natural gas emits 43% less carbon dioxide to produce the same amount of energy, albeit that estimate will vary depending on the type of coal.[230]

Natural gas liquifies at minus 162° Celsius (minus 260° Fahrenheit) and the cooling process shrinks the gas by a factor of 600. Since 1964, liquified natural gas has been transported on ships.[231] The liquefaction of natural gas is an example of an energy-related technology that completely transformed the relative competitiveness of a fuel. The abundance of natural gas in the United States has meant that a tremendous amount of low-cost, clean-burning energy is available to supply the American market and, once liquified, markets around the world.

---

[227] **Elements of Petroleum Geology** – Richard C. Selley and Stephen A. Sonnenberg – Elsevier – 2014
[228] **Units and calculators explained** – US Energy Information Administration – https://www.eia.gov/ – Accessed: August 2020
[229] **Natural Gas** – National Geographic Encyclopaedia – https://www.nationalgeographic.org/ – Accessed: March 2021
[230] **How much carbon dioxide is produced when different fuels are burned?** – US Energy Information Administration – https://www.eia.gov/ – Accessed: August 2020
[231] **Liquified Natural Gas (LNG)** – Shell – https://www.shell.com – Accessed: January 2021

Global energy demand from all sources increased by 2.6 times from 1971 to 2018, representing an average annual growth rate of 2.0% over that 47-year period.[232] Natural gas not only kept pace with growing energy demand, it increased its market share from 16% to 23%.[233] No other form of energy added 7% market share over that period. The runner up was nuclear energy, which added 4% market share.

The market share of natural gas for electricity generation in the United States overtook that of coal in 2016.[234] Adult Americans are currently enjoying the cleanest air of their lifetimes. Between 1970 and 2019 the combined emissions of the six most common pollutants dropped by 77%.[235] The increased use of natural gas has been a key driver of improved air quality – from blue flames, blue skies.

Natural gas and coal are not only useful for the purposes of creating fire. 6% of the natural gas supplied globally and 2% of the coal supplied globally is used as feedstock to produce hydrogen.[236] In turn, hydrogen is used mainly to create plant fertilizer, specifically, ammonia (NH_3).[237]

The world record crop yield for wheat was beaten three times in the five years up to 2020. It now stands at 17.398 metric tons of wheat from a hectare of land (2.5 acres).[238] In addition to providing fire, coal and natural gas both provide plant fertilizer, which is driving agricultural yields higher. Higher agricultural yields reduce the amount of land required for farming and thereby provide more land

---

[232] **World Energy Balances 2020 (Data for 2018)** – International Energy Agency – July 2020
[233] **World Energy Balances 2020 (Data for 2018)** – International Energy Agency – July 2020
[234] **Electricity in the United States** – US Energy Information Association – https://www.eia.gov – Accessed: January 2021
[235] **Our Nation's Air, Air Quality Improves as America Grows** – United States Environmental Protection Agency – https://gispub.epa.gov/air/trendsreport/2020 – Accessed: August 2020
[236] **The Future of Hydrogen** – The International Energy Agency, Technology Report – June 2019
[237] **Hydrogen in Industry** – Hydrogen Europe – https://hydrogeneurope.eu – Accessed: March 2021
[238] **Highest Wheat Yield** – Guinness World Records – https://www.guinnessworldrecords.com – Accessed: February 2021

for wildlife. Whether wildlife has land, or not, is *by far* the most important factor upon which the fortunes of wildlife hinge.

Fire from Bio-fuel, Farmed Fuel

Bio-fuels are blended with gasoline and diesel and burned in internal combustion engines.

In 2018, the world consumed 2.6 million barrels of liquid bio-fuel every day, representing an increase of 7% relative to the prior year.[239]

The two most important liquid bio-fuels are alcohol (ethanol), which is mixed into gasoline, and bio-diesel, which is mixed into diesel.

Alcohol for bio-fuel is derived from sweet and starchy crops such as corn and sugar cane. Bio-diesel is derived from organic fats and vegetable oil. The main crops that produce bio-diesel are palm oil and soybean.

Bio-fuels are associated with heavy social burdens – costs – and are uncompetitive relative to gasoline and diesel. Bio-fuels exist due to governmental interventions in markets.

The United States Environmental Protection Agency mandated that, in 2020, 20.09 billion gallons of bio-fuel must be blended into gasoline and diesel fuel.[240] That equates to about 11.0% of the Unites States' transportation fuel supply.[241]

The European Union mandated that by 2030 14% of the transport fuel consumed by its member countries must be from renewable sources, which it defines to include bio-fuels.[242]

In 2016, 413 thousand square kilometers (159 thousand square miles) of land was being used to grow food for the purposes of transforming

---

[239] **Renewables 2019** – The International Energy Agency – October 2019
[240] **Renewable Energy Fuel Standard, an Overview** – Congressional Research Service – April 14, 2020
[241] **US refiners required to blend 20.09 billion gallons renewable fuel in 2020: EPA** – Meghan Gordon – S&P Global, Platts – 19 December 2020
[242] **Renewable Energy** – Recast to 2030 (RED II) – European Commission – https://ec.europa.eu/ – Accessed: August 2020

it into transportation fuel.[243] For reference, that equates to 59% of the surface area of the state of Texas, USA.

Transport and Environment, a non-governmental organization, estimated that 53% of the European Union's bio-diesel is derived from imported crops.[244] The primary suppliers of European bio-diesel are poor countries located in the Tropics.[245] The green policies of the European Union are therefore a direct cause of large-scale deforestation in the Tropics.

Transport and Environment estimates that, net of all considerations, using bio-diesel rather than diesel actually *increases* carbon dioxide emissions by 80%.[246]

Alternatives to Fire

We will now look at forms of energy that are considered to have potential to replace the human use of fire. Specifically, we will look at hydro power, nuclear power, solar power and wind power. These sources of energy do not create atmospheric emissions of carbon dioxide, at least not directly. These sources of energy produce electricity. In 2018, electricity provided 19% of global energy.[247] In this section we will also assess means of storing electrical power.

Table 5 provides the sources of electrical energy globally.

---

[243] **The water-land-food nexus of first-generation biofuels** – Maria Cristina Rulli, Davide Bellomi, Andrea Cazzoli, Giulia De Carolis & Paolo D'Odorico – Scientific Reports – 3 March 2016
[244] **Around half of EU production of crop biodiesel is based on imports, not crops grown by EU farmers, new analysis** – Transport and Environment – 16 October 2017
[245] **The water-land-food nexus of first-generation biofuels** – Maria Cristina Rulli, Davide Bellomi, Andrea Cazzoli, Giulia De Carolis & Paolo D'Odorico – Scientific Reports – 3 March 2016
[246] **Globiom: the basis for biofuel policy post 2020** – Transport and Environment – April 2016
[247] **World Energy Outlook 2019** – International Energy Agency – November 2019

Sources of electrical energy — *Table 5*

Electricity by Source	
Coal	38.0%
Natural gas	23.0%
Hydro power	16.2%
Nuclear	10.1%
Wind	4.8%
Oil	2.9%
Biofuel and waste	2.4%
Solar	2.1%
Geothermal, tide and other	0.5%
Total	**100.0%**

Source: International Energy Agency 2020 (2018 data)[248]

In 2019, 770 million people did not have access to electricity.[249] The lower the cost of electricity, the more people will benefit from its use.

Hydro Power

Hydro power supplies 16.2% of electricity globally.[250]

In 1737, Bernard Forest de Bélidor of France published mathematical principles that created the foundations for the development of hydro power as we know it. In 1882, Nikola Tesla, who was born in Croatia before emigrating to the United States, invented an electrical generator capable of turning rotational power into electricity. That invention made it possible to transform hydro power into electricity.

Nuclear Power

Nuclear power supplies 10.1% of electricity globally.[251]

Nuclear power is premised on the theory of relativity as formulated by Albert Einstein. In 1905, he determined that energy is equivalent

---

[248] **Electricity Information Overview 2020 (data for 2018)** International Energy Agency – July 2020
[249] **Access to Electricity** – International Energy Agency – https://www.iea.org/ – Accessed: October 2020
[250] **Electricity Information Overview 2020 (data for 2018)** – International Energy Agency – July 2020
[251] **Electricity Information Overview 2020 (data for 2018)** – International Energy Agency – July 2020

to mass multiplied by the speed of light squared ($e = mc^2$). The implication of this formula is that mass can be converted into energy. Nuclear power plants exploit this by converting uranium into energy, principally thermal energy – heat. That heat is used to create steam, which is used to turn turbines that generate electricity.

The spent uranium fuel is radioactive and can remain so for thousands of years.[252]

Wind Power

Wind power provides 4.8% of electricity globally.[253] As a reminder, electricity provides 19% of global energy.[254] Wind power is widely considered to be a leading technology that could replace the human use of fire.

In 1887, James Blyth, a professor of Natural Philosophy, installed the first electricity-generating wind turbine in his hometown of Marykirk, Scotland. Using a lead-acid battery of French design, Professor Blyth charged his batteries when the wind was blowing for use anytime. Anecdotally, his offer to light up the main street of Marykirk was refused on the grounds that the new form of energy, electricity, was "the work of the devil."[255]

The principal challenges for wind energy are as follows:

i) Wind is an unreliable source of energy because wind speed is variable.

ii) Energy from wind creates high social burdens – costs. Positively, over a relatively short period spanning from 2010 to 2019, efficiency gains have reduced the costs of onshore and offshore wind energy by 39% and 29%, respectively.[256]

---

[252] **Radioactive Waste** – United States Nuclear Regulatory Commission, Office of Public Affairs – June 2019
[253] **Electricity Information Overview 2020 (data for 2018)** – International Energy Agency – July 2020
[254] **World Energy Outlook 2019** – International Energy Agency – November 2019
[255] **Papers of James Blyth** – Kirsteen Croll – University of Strathclyde Glasgow – https://www.strath.ac.uk – Accessed January 2021
[256] **Renewable Power Generation Costs in 2019** – International Renewable Energy Agency – 2020

iii) Wind turbines are intended to displace fire, but are dependent on materials derived from fire. The largest wind turbines today weigh in excess of 3,000 metric tons (excluding concrete and steel below the ground).[257] Wind turbines, by weight, consist of steel (69%-79%), iron (5%-17%), copper (1%), aluminum (0%-2%) and oil-derived materials such as plastic and fiber-glass (11%-16%).[258] As a reminder, it takes 0.77 tons of coal to produce one ton of raw steel.[259]

Wind turbines have grown in installed electricity-generating capacity by a factor of 75 times over the two decades to 2018.[260] In 2017, the International Energy Agency (IEA), commented on the source of the momentum in renewables indicating, "Globally the IEA estimates that $750 billion in economic incentives have been provided to renewables over the past decade." In addition to financial support, policies shelter wind producers from economic risks. Ørsted, the number-one ranked developer of offshore wind projects globally, stated, "Our offshore wind farms are largely subject to regulated prices, implying a high degree of revenue certainty."[261]

Solar Power

Solar power provides 2.1% of electricity globally.[262] As a reminder, electricity provides 19% of global energy.[263] Solar power is widely considered to be a leading technology that could replace the human use of fire.

In 1839, Frenchman Alexandre Edmond Becquerel arranged a metallic apparatus in an acidic solution. Unexpectedly, when the sun

---

[257] **Haliade-X uncovered: GE aims for 14MW** – Eize de Vries – Wind Power Monthly – 4 March 2019
[258] **2015 Cost of Wind Energy Review** – Christopher Mone, Maureen Hand, Mark Bolinger, Joseph Rand, Donna Heimiller and Jonathan Ho – National Renewable Energy Laboratory, United States Department of Energy – revised: May 2017
[259] **Metallurgical Coal** – BHP – https://www.bhp.com – Accessed: February 2021
[260] **Wind Energy** – International Renewable Energy Agency – https://www.irena.org – Accessed: October 2020
[261] **Ørsted Annual Report 2019** – https://orsted.com
[262] **Electricity Information Overview 2020 (data for 2018)** – International Energy Agency – July 2020
[263] **World Energy Outlook 2019** – International Energy Agency – November 2019

was shining on his apparatus, he observed that it created an electrical current. However, his apparatus had no commercial applications because it captured less than 1% of the energy in sunlight.

In the 1940s, Russell Shoemaker Ohl at Bell Laboratories in the United States noticed that a silicon semiconductor that had been accidentally cracked similarly produced an electrical current when exposed to sunlight – only this current was more powerful. Adaptations of this system captured about 1% of the energy in sunlight.[264]

In 2006, 13.2%-14.7% of the energy in sunlight was being captured by commercial solar panels. By 2019, that had risen to 17%-18% – an appreciable gain over a relatively short period of time.[265]

Solar power costs are falling at a rate of 85% per decade, representing a reduction in costs of 6.7 times per decade.[266]

By the 2050s, it is estimated that waste from solar panels will amount to 5.5 to 6.0 million metric tons per year.[267] Remarkably for an industry that is sustained by governmental regulations, the regulation of waste solar panels is just beginning to be addressed.[268]

Solar radiance is concentrated in the Tropics, making it the most ideal place on the planet for solar power (Figure 12).

---

[264] **April 25, 1954: Bell Labs Demonstrates the First Practical Silicon Solar Cell, This Month in Physics History** – APS News – April 2009
[265] **Future of Solar Photovoltaic** – International Renewable Energy Agency – November 2019.
[266] **Renewable Power Generation Costs in 2019** – International Renewable Energy Agency – 2020
[267] **End-of-Life-Management Solar Photovoltaic Panels** – International Renewable Energy Agency – June 2016
[268] **Solar Panels Information and FAQs** – Department of Toxic Substance Control, State of California – https://dtsc.ca.gov – Accessed: October 2020; and **If Solar Panels Are So Clean, Why Do They Produce So Much Toxic Waste?** – Michael Shellenberger – Forbes – 23 May 2018

Solar radiance by latitude — Figure 12

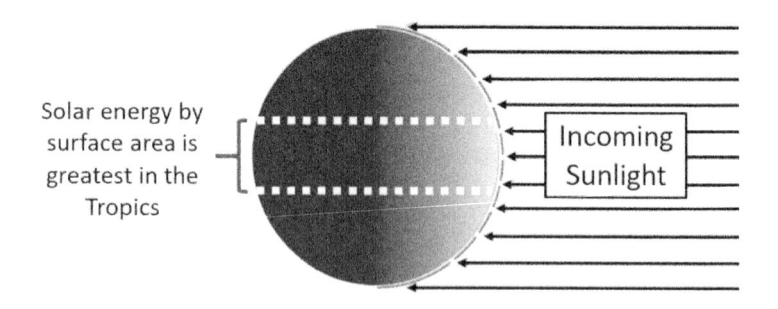

Solar energy by surface area is greatest in the Tropics

Incoming Sunlight

Image credit: Author

Solar power production in the Tropics has tremendous potential to create prosperity where it is needed most. Unlike bio-fuel production, which requires fertile land, solar power facilities can be located where vegetation and wildlife are least abundant.

High-resolution commercial satellites can provide images with a resolution of 30 centimeters (11.8 inches).[269] However, for the time being, computers are not sufficiently intelligent to determine using satellite data how much land is occupied globally by solar panels.[270]

Electrical Batteries

Fuels that burn, namely, coal, oil, natural gas and wood, represent both a source of energy and a store of energy. Electrical batteries represent a means of storing electricity produced from solar power and wind power. For reference, hydrogen also represents a potential store of electrical energy; hydrogen will be discussed in the next section.

In the last quarter of the 18th century, Luigi Galvani of Italy was able to induce a dissected frog's leg to twitch by touching it with bronze and iron metals while it was submerged in saltwater. Science was in

---

[269] **The Power of 30cm** – European Space Imaging – https://www.euspaceimaging.com – Accessed: March 2021

[270] **Distributed solar photovoltaic array location and extent dataset for remote sensing object identification** – Kyle Bradbury, Raghav Saboo, Timothy L. Johnson, Jordan M. Malof, Arjun Devarajan, Wuming Zhang, Leslie M. Collins & Richard G. Newell – Scientific Data – 6 December 2016

agreement that electricity was causing the frog's leg to twitch, but what was the source of the electricity? Allesandro Volta, also of Italy, argued that it came from the metals. In the process of proving his argument he invented the electrical battery.

Since Volta invented the battery, there have been two transformative developments in battery technology:

i) The first rechargeable battery was invented in 1859 by Gaston Planté of France. By reversing the current of electricity and sending it back into his batteries they would recharge.

ii) In 1991, Sony Corporation commercialized a revolutionary rechargeable battery concept based on the movement of lithium-ions within batteries.[271] Today, almost all mobile electric devices from smart phones to electric vehicles run on lithium-ion batteries.

Due to the unreliability of sunshine and wind, solar panels, onshore wind turbines and offshore wind turbines produce on average only 18%, 36% and 44%, respectively, of their maximum nameplate capacity.[272] As a result, and contrary to widely held perceptions, when those sources of energy are not available, charging battery powered electrical vehicles requires energy from the other sources shown in Table 5, namely, coal and natural gas.

For that reason and due to system inefficiencies, the Kiel Institute determined that, contrary to widely held perceptions, electric vehicles in Germany cause *more* greenhouse gas emissions than diesel-powered vehicles.[273] The International Energy Agency determined that globally the use of electric vehicles reduces carbon dioxide emissions, while not eliminating them.[274] The point is that when the sun is not shining nor the wind blowing, electricity is provided by other sources.

---

[271] **Camcorder to smartphone era, how lithium-ion batteries transformed tech** – Robert Masse – Business Standard – 16 July 2019
[272] **Renewable Power Generation Costs in 2019** – International Renewable Energy Agency – 2020
[273] **Electric Mobility and Climate Protection: A Substantial Miscalculation** – Ulrich Schmidt – Kiel Institute for the World Economy – June 2020
[274] **The Global EV Outlook 2020** – The International Energy Agency – June 2020

Electrical batteries have high social burdens – costs. Due to their high costs, they are used only where their advantages are greatest, namely, for mobile applications with frequent, short and regular charge-discharge cycles.

Today, commercial electrical batteries have the merit of being able to usefully discharge up to 96% of the energy used to charge them,[275] making them highly efficient on a round-trip basis. However, electrical batteries lose their charge when left idle. Electrical battery technology and performance are improving in almost all respects. Most importantly, according to the International Energy Agency, battery costs are falling at a rate of about 89%, or by 8.8 times, per decade.[276] New applications for batteries will result from their falling costs.

Increasingly, the environmental costs of sourcing materials from developing countries to manufacture electrical batteries are being scrutinized.[277] Currently, there are no obvious means of safely recycling lithium-ion batteries after their useful lives.[278]

Hydrogen

Hydrogen, like electrical batteries, represents a means of storing energy.

Hydrogen can be produced by splitting water (H_2O) into oxygen (O_2) and hydrogen (H_2) using electricity in a process called electrolysis. Today, less than 0.1% of hydrogen is produced from electrolysis.[279] Hydrogen from electrolysis represents a means of storing energy produced from wind power and solar power.

---

[275] **Product datasheet: Intesium® Max+ 20M** – SAFT – https://www.saftbatteries.com – Accessed: December 2020

[276] **The Global EV Outlook 2020** – The International Energy Agency – June 2020

[277] **Developing countries pay environmental cost of electric car batteries** – United Nations Conference on Trade and Development (UNCTAD) – 22 July 2020 (based on: Commodities at a Glance No 13, Special issue on strategic battery raw materials – UNCTAD – 2020)

[278] **Recycling lithium-ion batteries from electric vehicles** – Gavin Harper, Roberto Sommerville, Emma Kendrick, Laura Driscoll, Peter Slater, Rustam Stolkin, Allan Walton, Paul Christensen, Oliver Heidrich, Si-mon Lambert, Andrew Abbott, Karl Ryder, Linda Gaines and Paul Anderson – Nature – 6 November 2019

[279] **The Future of Hydrogen** – International Energy Agency – June 2019

Hydrogen can be burned. However, hydrogen produced from electrolysis is intended to provide the energy to power fuel cells. Fuel cells produce electricity directly without a need for fire. Hydrogen, whether burned or consumed in a fuel cell, emits only water as a by-product. Consuming hydrogen does not emit carbon dioxide.

Hydrogen has the advantage of providing 2.7 times more energy than gasoline by weight.[280] However, it is voluminous; even when compressed to 700 times atmospheric pressure it requires 5.7 times more volume than gasoline for the same amount of energy.[281] Compressing hydrogen consumes an amount of energy equivalent to 12%-15% of the energy so compressed.[282]

Unlike the energy in electrical batteries, the energy stored in hydrogen remains constant over time – a critical advantage. However, currently, using hydrogen to store electricity loses about 64% of the initial energy in the process of converting electricity into hydrogen and back into electricity.[283] This increases the cost of energy stored in hydrogen by 2.7 times relative to the cost of the initial electrical energy (ignoring equipment and compression costs). However, reliable energy stored in the form of hydrogen is immeasurably more valuable than energy that is available only when the wind is blowing or the sun is shining. Positively, the round-trip efficiency of storing electricity in hydrogen is expected to increase significantly with technological advancements.[284] Most importantly, the cost of hydrogen derived from wind power and solar power will also fall with rapidly falling wind power and solar power costs. If the costs of solar power continue to fall at their current pace, hydrogen sourced from solar power will inevitably become the most cost-competitive form of energy that can be stored indefinitely for reliable use anytime.

---

[280] **Hydrogen Storage** – US Office of Energy Efficiency and Renewable Energy – https://www.energy.gov/ – Accessed: August 2020
[281] **Hydrogen - A sustainable energy carrier** – Kasper T. Møller, Torben R. Jensen, Etsuo Akiba and Hai-wen Li – Progress in Natural Science: Materials International – 27(2017)
[282] **Green Hydrogen in Developing Countries** – World Bank, ESMAP – 2020
[283] **Product datasheet: HgasXMW Product Specification** – ITM Power – https://www.itm-power.com – Accessed: October 2020; and
Green Hydrogen in Developing Countries – World Bank, ESMAP – 2020
[284] **Solid Oxide Electrolysis** – Ceres Power – Final Results – 17 March 2021

Costs of Fire vs. Alternatives to Fire

The most critically misunderstood consideration in respect of wind power and solar power is as follows: To assess the social burdens – costs – of wind and solar power as a means of providing reliable energy requires the inclusion of the costs of storing energy from those sources.

In the absence of a means of storing electrical energy, the costs of wind and solar power are meaningless because those sources of energy would require a duplication of investment in backup power for when they are unavailable.

Table 6 provides an estimate of the costs of reliable electrical energy derived from wind power and solar power that has been stored in electrical batteries or in the form of hydrogen. For reference, all cost estimates in relation to wind power, solar power and electrical battery storage have been provided by the International Renewable Energy Agency.

Table 6 also provides the costs of oil, natural gas and coal, based on their average prices in 2019. A price reference for nuclear power has also been provided based on the Hinkley Point C nuclear power plant, which is under construction in the United Kingdom.

Costs of fire vs. alternatives to fire — Table 6

	Price 2019 Average ($/MWh)	Costs of Reliable Electricity from Alternatives to Fire (Wind & Solar Power) As Stored In	
		Electrical Batteries ($/MWh)	Hydrogen ($/MWh)
Onshore wind power	53.0	n.a.	**145.6**
Offshore wind power	115.0	n.a.	**315.9**
Solar power	68.0	**176-208**	**186.8**
Oil ($64.00/bbl; Brent)	**34.1**	n.a.	n.a.
US Coal ($38.53/delivered ton)	**7.2**	n.a.	n.a.
US natural gas ($2.57/mmbtu)	**8.8**	n.a.	n.a.
UK nuclear (£92.5/MWh)	**118.1**	n.a.	n.a.

Sources[285]

Currently, fire, namely, fire from oil, coal and natural gas, has *by far* the lowest social burdens – costs – relative to the three leading alternatives to fire, namely, nuclear power, wind power and solar power (Table 6). Reliable alternatives to fire are in many cases more than 10 times more burdensome on society – costly – than energy from fire (Table 6).

For reference, the costs in Table 6 can be considered to reflect average costs as estimated in 2019. Some projects are better than average and some worse.

---

[285] **Energy conversion calculators** – US Energy Information Administration – https://www.eia.gov – Accessed: October 2020;
Renewable Power Generation Costs in 2019 – International Renewable Energy Agency – 2020;
Utility-scale Batteries, Innovation Landscape Brief – Arina Anisie and Francisco Boshell – International Renewable Energy Agency – 2019;
Electrical Energy to Hydrogen Conversion Efficiency from HgasXMW Product Specification – ITM Power – https://www.itm-power.com – Accessed: October 2020;
Hydrogen to Electrical Energy Conversion Efficiency from Green Hydrogen in Developing Countries – World Bank, ESMAP – 2020;
US energy prices – US Energy Information Administration – https://www.eia.gov – Accessed: December 2020; and
Hinkley Point C - UK BEIS – https://www.gov.uk – Accessed: December 2020

Let us gain perspective by assessing how much of our energy is actually provided by fire and how much is provided by alternatives to fire.

Global Energy Supply by Source

Contrary to prevailing perceptions, fire, which includes fire from oil, coal, natural gas and bio-fuels, still accounts for 90.5% of all energy production globally, according to the International Energy Agency.

That figure is 85.0% in the United States, according to the US Energy Information Administration and 81.5% in the European Union, according to the International Energy Agency.

A summary of the sources of energy from a wholistic perspective for the world, the United States and the European Union is provided in Table 7.

Energy supply by source — Table 7

	World	**USA**	**EU**
Oil	31.5%	36.7%	32.8%
Natural Gas	22.8%	32.1%	24.6%
Coal	26.9%	11.3%	14.3%
Biofuels	9.3%	4.9%	9.8%
Total from Fire	**90.5%**	**85.0%**	**81.5%**
Nuclear	4.9%	8.5%	13.4%
Hydro	2.5%	2.5%	1.6%
Wind		2.8%	1.9%
Solar	2.1%	1.0%	1.0%
Other		0.1%	0.6%
Total	**100.0%**	**100.0%**	**100.0%**
Year of data	*2018*	*2019*	*2017*
Year of publication	*2020*	*2020*	*2020*
Source	*IEA*	*EIA*	*IEA*

Sources: International Energy Agency, US Energy Information Administration[286]

It is important to appreciate that energy statistics that show the market share of wind power and solar power relative to the electricity

---

[286] **World Energy Balances 2020 (Data for 2018)** – International Energy Agency – July 2020;

market will increase the apparent contributions of those forms of energy by a factor of about five times. This is because electricity represents only about one fifth of the energy market.[287] It is also important to be aware that presenting "renewables" as a single category that includes either hydro power or bio-fuels can substantially increase the apparent contributions made by wind power and solar power.[288]

US primary energy consumption by energy source, 2019 – US Energy Information Administration – https://www.eia.gov – Accessed: January 2021; and
European Union 2020, Energy Policy Review (Data for 2017) – International Energy Agency
[287] **World Energy Outlook 2019** – International Energy Agency – November 2019
[288] **Energy balance sheets, 2020 Edition** – Eurostat – 2020

Part 3: Extinguishing the Human Use of Fire

1. From Fire: Pollution

Fire has brought countless benefits, but it is also a major cause of air pollution.

On average, air pollution globally is estimated to shorten each human life by 20 months.[289] Air pollution is ranked 4th globally as a cause of death, behind high blood pressure, tobacco and diet.[290] The World Health Organization suggests that roughly half of the deaths caused by air pollution are related to ambient air pollution, with the other half related to household pollution, mostly from burning solid fuels for cooking.[291] Positively, cleaner air trends globally would suggest that going forwards we should expect to live both cleaner and longer lives.

A key indicator of air quality is the extent to which air is free of fine particle pollution.[292] Fine particle pollution is defined as airborne particle matter less than 2.5 microns in size. For perspective, human hair has a width of about 70 microns.

Fine particle pollution is particularly dangerous because it enters deep into people's lungs and can enter into their bloodstream.[293] Air with more than 10 micrograms per cubic meter (10 µg/m^3) of fine particle pollution poses a health risk.[294]

For the average person globally, the level of fine particle pollution peaked in 2014 at 47.4 µg/m^3 and fell to 42.6 µg/m^3 in 2019.[295]

In 2019, the ten countries with the cleanest air as measured by fine particle pollution were: Australia, Brunei, Canada, Estonia, Finland, Iceland, New Zealand, Norway, Sweden and the United States. These

---

[289] **State of Global Air/2019** – Health Effects Institute – 2019
[290] **State of Global Air/2020** – Health Effects Institute – 2020
[291] **Air Pollution** – World Health Organization – https://www.who.int/ – Accessed: October 2020
[292] **State of Global Air/2020** – Health Effects Institute – 2020
[293] **Particle Pollution** – Centers for Disease Control and Prevention – https://www.cdc.gov – Accessed: October 2020
[294] **Ambient (outdoor) air pollution** – World Health Organization – https://www.who.int – Accessed: October 2020
[295] **State of Global Air/2020** – Health Effects Institute – 2020

countries have average fine particle pollution levels of 8μg/m^3 or less.[296]

Focusing on Europe's top three economies, in 2019, fine particle pollution in Germany, the United Kingdom and France amounted to 11.8μg/m^3, 10.1μg/m^3 and 11.4μg/m^3, respectively.[297]

In contrast, the least developed countries suffer the worst air quality.[298]

Burning coal in power plants and burning oil-derived products in automobiles contributes to higher levels of fine particle pollution, amongst other air pollutants.

Focusing on the United States, compared to in the 1970s, vehicles today are over 75% more powerful, 90% more fuel efficient and 99% cleaner (for common pollutants: hydrocarbons, carbon monoxide, nitrogen oxides and particle emissions).[299] This highlights that i) air pollution and performance are unrelated and ii) the few worst polluters, for example old vehicles, can be 100 times worse than average.

Although many global organizations suggest that reducing the use of fire from coal, oil and natural gas will improve air quality, this overlooks that i) fire from coal, oil and natural gas has potential to increase prosperity, which is strongly associated with clean air and ii) the growth of natural gas consumption has been a primary driver of cleaner air.

The trajectory of declining air pollution is extremely encouraging and there is every reason to expect this trend to continue. The long-term fortunes of coal, oil and natural gas will likely depend on the degree to which they contribute to efforts to ensure the quality of our air continues to improve. Likewise, the heavy reliance of wind turbines on steel, and therefore on coal, which has potential to cause significant pollution, will likely weigh on the long-term outlook for that source of energy.

---

[296] **State of Global Air/2020** – Health Effects Institute – 2020
[297] **State of Global Air** – https://www.stateofglobalair.org – Accessed: October 2020
[298] **State of Global Air/2019** – Health Effects Institute – 2019
[299] **Highlights of the Automotive Trends Report** – United States Environmental Protection Agency – https://www.epa.gov – Accessed: October 2020

Efforts to reduce air pollution have been highly successful decade-over-decade and have had limited social burdens – costs.

In contrast, despite the extremely heavy social burdens – costs – incurred to reduce the rise in atmospheric levels of carbon dioxide,[300] these efforts have had no notable impact (Figure 13). For reference, carbon dioxide emissions set a new record high in 2019.[301]

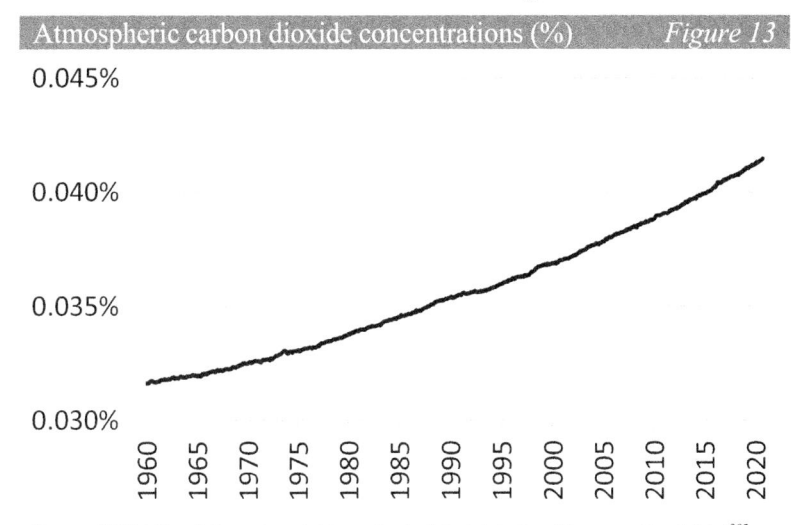

Atmospheric carbon dioxide concentrations (%) — *Figure 13*

Source: US National Oceanic and Atmospheric Administration (deseasonalized data)[302]

---

[300] **The Success of Wind and Solar is Powered by Strong Policy Support** – Laura Cozzi, Tim Gould and Paolo Frankl – The International Energy Agency – 1 June 2017
[301] **Analysis: Global fossil-fuel emissions up 0.6% in 2019 due to China** – Zeke Hausfather – Carbon Brief – https://www.carbonbrief.org/ – Accessed: October 2020
[302] **Monthly Average Mauna Loa CO_2** – US National Oceanographic and Atmospheric Administration – https://www.esrl.noaa.gov – Accessed: August 2020

2. From Fire: Global Warming

According to NASA, the planet's average surface temperature has risen by 1.18° Celsius (2.12° Fahrenheit) since the late 1800s.[303] NASA estimates that the rate of temperature increase in the last century was "roughly ten times faster than the average rate of ice-age-recovery warming" related to our continued withdrawal from the last period of glaciation.[304]

In this section the global warming paradigm that has been accepted since 1990 to explain the Earth's rising surface temperatures will be discussed. That paradigm forms the basis of the justification to eliminate the human use of fire.

Let us begin by looking at thermal radiation. Thermal radiation is emitted by all things that have a temperature above minus 273° Celsius (minus 460° Fahrenheit; 0° Kelvin), which is defined as absolute zero. Above that temperature the energy in matter causes all particles to vibrate, which results in the emission of thermal radiation. The Sun emits thermal radiation because it is hot. Likewise, the Earth emits thermal radiation too because it is hotter than absolute zero. The most well-known form of thermal radiation is light because it is visible. The thermal radiation emitted by the Earth is long-wave thermal radiation and it is not visible to the naked eye. Matter emits thermal radiation across a spectrum of wavelengths, not just a single wavelength.

In 1859, John Tyndall discovered that some gases let thermal radiation pass through them and that other gases absorb thermal radiation. Closer to his own words, he determined that there were differences in the abilities of "perfectly colorless and invisible gases and vapors" to absorb thermal radiation.[305]

---

[303] **Climate Change: How Do We Know?** – Global Climate Change, NASA – https://climate.nasa.gov – Accessed: February 2021
[304] **How is Today's Warming Different from the Past?** – NASA Earth Observatory – https://Earthobservatory.nasa.gov – Accessed: September 2020
[305] **John Tyndall (1820-1893)** – Steve Graham – NASA Earth Observatory – 8 October 1999 – https://Earthobservatory.nasa.gov/ – Accessed: September 2020

Several gases in our atmosphere such as water vapor (H_2O), carbon dioxide (CO_2), oxygen (O_2), ozone (O_3), methane (CH_4), nitrogen oxide (N_2O) and carbon monoxide (CO) absorb thermal radiation to varying degrees depending on the thermal radiation's wavelengths. In contrast, certain gases do not absorb thermal radiation. Nitrogen (N_2) constitutes 78.1% of our atmosphere and it does not absorb thermal radiation.

John Tyndall invented an apparatus to measure the absorptive powers of various gases called a spectrophotometer. With this instrument it is possible to determine which wavelengths of thermal radiation get absorbed by various gases.

The percentage of thermal radiation leaving the Earth that is absorbed by various gases in the atmosphere across a range of wavelengths is shown in Figure 14.

Figure 14 also shows the range of thermal radiation that would be emitted by the Earth under the theoretical assumption that it has a uniform temperature of 20° Celsius (68° Fahrenheit).

Combined, Figure 14 shows that a significant amount of the thermal radiation leaving the Earth is absorbed by greenhouse gases rather than being radiated directly out to space.

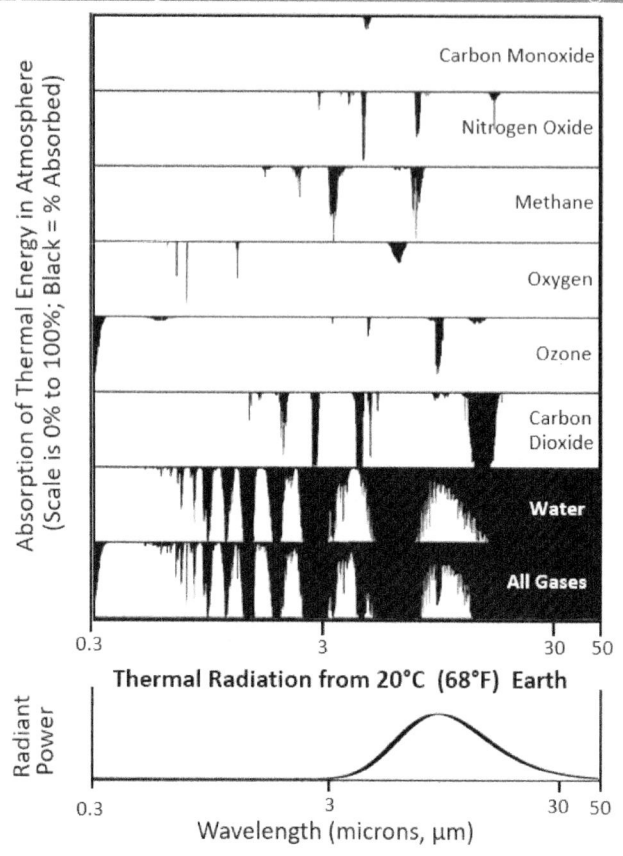

Atmospheric absorption of thermal radiation — *Figure 14*

Source: Author (adapted from various[306])

For reference, Figure 14 is intended only for indicative purposes, it is an oversimplification of the complexity of our Earth and atmosphere.

---

[306] **Adapted by Author from: Handbook of Geophysics and Space Environments** – Shea L. Valley – McGraw Hill – 1965;
Radiation Transmitted by the Atmosphere – Wikipedia (Accessed: September 2020);
Global Warming Art – Robert Rohde – via Climate Forcings and Global Warming – NASA Earth Observatory – https://Earthobservatory.nasa.gov – Accessed: September 2020); and
Black Body Radiation Equation – Courtesy of Professor Kurt Hollocher – Geology Department, Union College, New York

When a gas molecule absorbs thermal radiation, it re-radiates that thermal radiation in all directions until the energy dissipates. As a result, thermal radiation absorbed by gases in the atmosphere is partially re-radiated back to Earth. This increases the surface temperature of the Earth.

John Tyndall's discovery explained what has become known as the greenhouse effect. The heat-absorbing gases in our atmosphere have become known as greenhouse gases.

John Tyndall concluded that amongst the greenhouse gases water vapor is the strongest absorber of thermal radiation and is therefore the most important gas controlling the Earth's surface temperature. He determined that without water vapor, the Earth's surface would be "held fast in the iron grip of frost."[307]

This phenomenon was understood long before John Tyndall explained its root cause. Humid nights retain the Earth's heat. In contrast, when the air is dry, during the night, the Earth cools quickly.

It has been estimated that without greenhouse gases in our atmosphere the Earth would be 33° Celsius (59° Fahrenheit) colder.[308] The relative contributions of atmospheric constituents to this 33° Celsius (59° Fahrenheit) increase in the Earth's surface temperature are as follows: water vapor and clouds: 75%; carbon dioxide: 20%; and other constituents: 5%.[309]

In 1896, Svante Arrhenius, a Swedish scientist with a background in electrochemistry, became the first person to suggest that carbon dioxide from burning coal would increase the surface temperature of the Earth.[310] The conceptual basis for this assertion was that

---

[307] **John Tyndall** – Steve Graham – NASA Earth Observatory – 8 October 1999 – https://Earthobservatory.nasa.gov – Accessed: September 2020

[308] **Climate Change, the IPCC Scientific Assessment** – World Meteorological Organization/United Nations Environment Programme, Intergovernmental Panel on Climate Change – Editors: J.T. Houghton, G.J. Jenkins and J.J. Ephraums – Cambridge University Press – 1990

[309] **Attribution of the present-day total greenhouse effect** – Gavin A. Schmidt, A. Ruedy. Ron L. Miller, Andy A. Lacis – Journal of Geophysical Research – 16 October 2010

[310] **On the Influence of Carbonic Acid in the Air upon the Temperature of the Ground** – Svante Arrhenius – Philosophical Magazine and Journal of Science – April 1896

increased carbon dioxide levels in the atmosphere would increase the amount of thermal radiation re-radiated back to Earth.

Guy Calendar, who was born in Canada but lived as an adult in England, collected temperature measurements and determined, in 1938, that the Earth had warmed and further suggested that this was the result of human emissions of carbon dioxide.[311] Both Svante Arrhenius and Guy Calendar developed mathematical models to develop their conclusions.

The science of how the Earth is changing due to increased atmospheric concentrations of carbon dioxide exists in mathematical models – today, computer models. The most recent forward-looking temperature forecasts of the Intergovernmental Panel on Climate Change for their "high CO_2 scenario" is the average forecast of 39 computer models.[312]

Computer models have determined that rising carbon dioxide levels increase the surface temperature of the Earth through the greenhouse effect and in doing so cause 20% of global warming. Other greenhouse gases emitted by humans, such as methane, cause an additional 5% of global warming.[313]

The remaining 75% of global warming is caused by feedback effects according to computer models.[314] According to computer models, increased surface temperatures caused by rising carbon dioxide levels kick-start a warming process that further increases the surface

---

[311] **The artificial production of carbon dioxide and its influence on temperature** – G.S. Calendar – Quarterly Journal of the Royal Meteorological Society – April 1938
[312] **Climate Change 2014, Synthesis Report** – Editors: Rajendra K. Pachauri, Leo Meyer and Core Writing Team – Fifth Assessment Report of the Intergovernmental Panel on Climate Change – 2015
[313] **CO2: The Thermostat that Controls Earth's Temperature** – Andrew Lacis – NASA Goddard Institute for Space Studies – October 2020 –
https://www.giss.nasa.gov/ – Accessed: September 2020; and
Attribution of the present-day total greenhouse effect – Gavin A. Schmidt, A. Ruedy. Ron L. Miller, Andy A. Lacis – Journal of Geophysical Research – 16 October 2010
[314] **CO2: The Thermostat that Controls Earth's Temperature** – Andrew Lacis – NASA Goddard Institute for Space Studies – October 2020 –
https://www.giss.nasa.gov/ – Accessed: September 2020; and
Attribution of the present-day total greenhouse effect – Gavin A. Schmidt, A. Ruedy. Ron L. Miller, Andy A. Lacis – Journal of Geophysical Research – 16 October 2010

temperatures of the Earth. Specifically, higher surface temperatures are assumed to increase the amount of water vapor in the atmosphere. More water vapor in the atmosphere further amplifies the greenhouse effect because water vapor is *by far* the most important greenhouse gas. This process underlines the "feedback" effects resulting from rising levels of carbon dioxide in the atmosphere.

By its construction, the global warming paradigm that has existed since 1990 specifies that carbon dioxide emissions from fire, and other greenhouse gases emitted by humans, are the only cause of changes to our climate.

According to the global warming paradigm that has existed since 1990, changes to the amount of water vapor in the atmosphere are caused only by heat-related feedback effects – nothing else. Citing from the United Nation's Intergovernmental Panel on Climate Change, the concentration of water vapor in the atmosphere is "determined internally within the climate system."[315]

In the next section we will discuss the recent changes that have occurred related to the science of global warming.

---

[315] **Climate Change, the IPCC Scientific Assessment** – United Nations Intergovernmental Panel on Climate Change – 1990

3. From Fire: Carbon Dioxide Fertilization

Launched in 1999, NASA's Terra satellite has just been in orbit long enough to provide meaningful data showing how the Earth is changing over time. Since 2016, publications revealing how our Earth is actually changing based on satellite data have been completely surprising and contrary to expectations. As a result, our understandings of the Earth and of climate science are changing.

Climate science has also been transformed over the last decade by the acceptance that carbon dioxide fertilizes plant growth and that trees and vegetation emit considerably more water vapor than previously understood.

Carbon Dioxide's Fertilization Effect: In 2016, a team of 32 authors from 24 globally recognized institutions identified carbon dioxide as being the primary driver causing our Earth to become greener through the fertilization of plant growth.[316]

Carbon dioxide fertilization explains 70% of the observed greening trend on our Earth, followed by nitrogen deposition (9%), climate change (8%) and land cover change (4%).[317]

Based on these attributions, carbon dioxide emitted by fire, namely, fire from burning coal, oil and natural gas, is the driving force that is making our Earth greener.

Forested Planet: In 2018, a team of scientists, including many from NASA itself, published a report, based on satellite data, indicating that "contrary to the prevailing view that forest area has declined globally," the Earth's forested surface area increased by 2.24 million square kilometers (864 thousand square miles), net of losses, over the

---

[316] **Carbon Dioxide Fertilization Greening Earth, Study Finds** – NASA – https://www.nasa.gov – Accessed: March 2021; and
Greening of the Earth and its Drivers – Zaichun Zhu and 31 additional authors – Nature Climate Change – 25 April 2016
[317] **Greening of the Earth and its Drivers** – Zaichun Zhu and 31 additional authors – Nature Climate Change – 25 April 2016

35 years to 2016.[318] That change represents a 2.0% increase in the Earth's forested surface area, net of losses, per decade.

The rates at which our Earth is increasing in forested surface area, net of losses, over various measures of time and surface area are provided in Table 8.

Rate of increase in the Earth's forested surface area — Table 8

	American Football Fields	Square Kilometers	Square Miles
Per year	11,922,642.2	64,000.0	24,710.5
Per day	32,644.3	175.2	67.7
Per hour	1,360.2	7.3	2.8
Per minute	22.7	0.1	0.0
Per second	0.4	0.0	0.0

Source[319]

The rate of growth in the Earth's forested surface area, net of losses, equates to 92% of the surface area of the state of Texas, USA, every ten years.

As an important digression, in contradiction to the satellite data, the United Nations reported in 2020 that, "In absolute terms, the global forest area *decreased* by 1.78 million square kilometers (0.69 million square miles) between 1990 and 2020."[320] What is going on?

The United Nations has not recognized the satellite data. It defines forests as a "land use" category.[321] As a result, whether a forest grows or whether it has trees at all is not of primary relevance for the United Nations. The data of the United Nations is for socio-economic analysis, not for science. It does, however, indicate the critical

---

[318] **Global land change from 1982 to 2016** – Xiao-Peng Song, Matthew C. Hansen, Stephen V. Stehman, Peter V. Potapov, Alexandra Tyukavina, Eric F. Vermote & John R. Townshend – Nature – 8 August 2018
[319] **Global land change from 1982 to 2016** – Xiao-Peng Song, Matthew C. Hansen, Stephen V. Stehman, Peter V. Potapov, Alexandra Tyukavina, Eric F. Vermote & John R. Townshend – Nature – 8 August 2018
[320] **The State of the World's Forests 2020** – United Nations Food and Agricultural Organization – 2020
[321] **Forest vs. Tree Cover: What is the Difference, The State of the World's Forests 2020** – United Nations Food and Agricultural Organization – 2020

importance of having high-quality unbiased satellite data from trustworthy sources.

In the future, science can be expected to align its understandings with satellite data.

Green Planet: In 2019, satellite data from NASA's Terra satellite was published indicating that the green leaf area of the Earth increased by 5.4 million square kilometers (2.1 million square miles) over the 18 years to 2017.[322] That represents an increase of 2.3% in the Earth's leaf area, net of losses, per decade.

The expansion of the green area of the Earth is 4.7 times greater than the expansion of forests.

One third of all vegetated land on Earth exhibited greening and only 5% exhibited browning over the period of study.[323]

It is interesting to contrast the satellite data with the assumption upon which the global warming paradigm has been constructed: According to the United Nation's Intergovernmental Panel on Climate Change, "There are no field data from whole ecosystem studies of forests that demonstrate a CO_2 fertilization effect."[324]

The rates at which our Earth is greening over various measures of time and surface area are provided in Table 9.

---

[322] **China and India Lead the Way in Greening** – Abby Tabor – NASA Earth Observatory – https://earthobservatory.nasa.gov – Accessed: March 2021; and
China and India lead in greening of the world through land-use management – Chi Chen, Taejin Park, Xuhui Wang, Shilong Piao, Baodong Xu, Rajiv K. Chaturvedi, Richard Fuchs, Victor Brovkin, Philippe Ciais, Rasmus Fensholt, Hans Tømmervik, Govindasamy Bala, Zaichun Zhu, Ramakrishna R. Nemani & Ranga B. Myneni – Nature Sustainability – 11 February 2019

[323] **China and India lead in greening of the world through land-use management** – Chi Chen, Taejin Park, Xuhui Wang, Shilong Piao, Baodong Xu, Rajiv K. Chaturvedi, Richard Fuchs, Victor Brovkin, Philippe Ciais, Rasmus Fensholt, Hans Tømmervik, Govindasamy Bala, Zaichun Zhu, Ramakrishna R. Nemani & Ranga B. Myneni – Nature Sustainability – 11 February 2019

[324] **Climate Change, the IPCC Scientific Assessment** – United Nations Intergovernmental Panel on Climate Change – 1990

Rate of increase in the Earth's green surface area — Table 9

	American Football Fields	Square Kilometers	Square Miles
Per year	55,887,385.3	300,000.0	115,830.6
Per day	153,023.0	821.4	317.2
Per hour	6,376.0	34.2	13.2
Per minute	106.3	0.6	0.2
Per second	1.8	0.0	0.0

Source[325]

The rate of growth in the Earth's newly green surface area equates to 37% of the surface area of the Lower-48 States of the United States every ten years.

Trees are water fountains: In the last decade, science has acknowledged that trees are water fountains in terms of the effect they have on the amount of water vapor in the atmosphere.[326]

On average the atmosphere globally contains the equivalent of 2.5 centimeters (1.0 inches) of water in the form of vapor.[327] By weight 1% of the atmosphere consists of water.[328]

As a reminder, 75% of the greenhouse gas effect[329] and 75% of global warming[330] are attributed to water vapor and clouds.

---

[325] **China and India lead in greening of the world through land-use management** – Chi Chen, Taejin Park, Xuhui Wang, Shilong Piao, Baodong Xu, Rajiv K. Chaturvedi, Richard Fuchs, Victor Brovkin, Philippe Ciais, Rasmus Fensholt, Hans Tømmervik, Govindasamy Bala, Zaichun Zhu, Ramakrishna R. Nemani & Ranga B. Myneni – Nature Sustainability – 11 February 2019

[326] **The largest river on Earth is invisible** — and airborne – Dan Kedmey – Science – 24 November 2015; and
Study shows the Amazon makes its own rainy season – Carol Rasmussen – NASA's Jet Propulsion Laboratory – 17 July 2017

[327] **The Atmosphere and the Water Cycle** – US Geological Survey – https://www.usgs.gov – Accessed: March 2021

[328] **Forests, atmospheric water and an uncertain future: the new biology of the global water cycle** – Douglas Sheil – Forest Ecosystems – 20 March 2018

[329] **Attribution of the present-day total greenhouse effect** – Gavin A. Schmidt, A. Ruedy, Ron L. Miller, Andy A. Lacis – Journal of Geophysical Research – 16 October 2010

[330] **CO2: The Thermostat that Controls Earth's Temperature** – Andrew Lacis – NASA Goddard Institute for Space Studies – October 2020 – https://www.giss.nasa.gov/ – Accessed: September 2020; and

It is now recognized that forests typically emit more water vapor than open water.[331] In 2009, it was estimated that the Amazon rainforest emits water vapor amounting to 1.37 meters (4.5 feet) of water per year into the atmosphere – rain in reverse.[332] Similar estimates for forests elsewhere in the world are currently being established.

In 2017, it was estimated that 61% of the rain that falls on land is derived from land evaporation and that only 39% is derived from oceanic evaporation.[333] Also in 2017, a team of international experts determined that globally 57.2% of land-derived water vapor is from vegetation.[334]

In 2011, moisture evaporated from land was estimated to typically travel 500 to 5,000 kilometers (300 to 3,000 miles) before returning to the Earth, with recycle times of 3 to 20 days.[335]

Although these critical estimates are evolving, their conceptual implications are obvious: Trees emit colossal amounts of the most important greenhouse gas, water.

Based on satellite data, the amount of water vapor being emitted by vegetation into the atmosphere is rising every decade as a result of the Earth's increasing green and forested surface areas.[336]

Attribution of the present-day total greenhouse effect – Gavin A. Schmidt, A. Ruedy. Ron L. Miller, Andy A. Lacis – Journal of Geophysical Research – 16 October 2010
[331] **Forests, atmospheric water and an uncertain future: the new biology of the global water cycle** – Douglas Sheil – Forest Ecosystems – 20 March 2018
[332] **The land–atmosphere water flux in the tropics** – Fisher JB, Malhi Y, Bonal D, Da Rocha HR, De Araújo AC, Gamo M, Goulden ML, Rano TH, Huete AR, Kondo H, Kumagai T, Loescher HW, Miller S, Nobre AD, Nouvellon Y, Oberbauer SF, Panuthai S, Roupsard O, Saleska S, Tanaka K – Global Change Biology – (15) 2009
[333] **Evaluating the Hydrological Cycle over Land Using the Newly-Corrected Precipitation Climatology from the Global Precipitation Climatology Centre** – Udo Schneider, Peter Finger, Anja Meyer-Christoffer, Elke Rustemeier, Markus Ziese and Andreas Becker – Atmosphere – 3 March 2017
[334] **Revisiting the contribution of transpiration to global terrestrial evapotranspiration** – Zhongwang Wei, Kei Yoshimura, Lixin Wang, Diego G. Miralles, Scott Jasechko and Xuhui Lee – Geophysical Research Letters (AGU Publications) – 31 March 2017
[335] **Length and time scales of atmospheric moisture recycling** – R. J. van der Ent and H. H. G. Savenije – Atmospheric Chemistry and Physics – 1 March 2011
[336] **Responses of land evapotranspiration to Earth's greening in CMIP5 Earth System Models** – Zhenzhong Zeng, Zaichun Zhu, Xu Lian, Laurent Z X Li, Anping

A forest in Poland (latitude 49° North) interacting with the atmosphere is shown in Figure 15.

Every tree is a water fountain *Figure 15*

Photo Credit: Marek Piwnicki on Unsplash (Location: Bieszczady Mountains, Poland)

Forests Absorb Heat: Not only do forests emit water vapor, the most important greenhouse gas, forests absorb more (reflect less) thermal radiation from the Sun, this is shown in Figure 16.

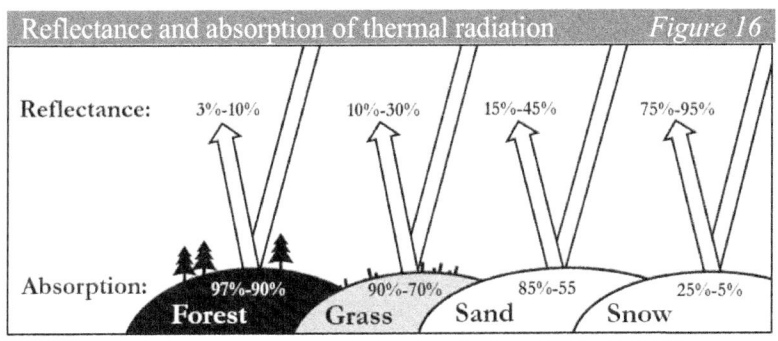

Reflectance and absorption of thermal radiation *Figure 16*

Reflectance: 3%-10% | 10%-30% | 15%-45% | 75%-95%

Absorption: 97%-90% | 90%-70% | 85%-55 | 25%-5%
Forest | **Grass** | **Sand** | **Snow**

Image Credit: Author; data source: Goddard Institute for Space Studies NASA[337]

Northern Hemisphere Warming: Most of the newly green and forested surface area of the Earth is in the northern hemisphere where

---

Chen, Xiaogang He and Shilong Piao – Environmental Research Letters – 3 October 2016

[337] **NASA Goddard Institute for Space Studies (GISS) Climate Change Research Initiative (CCRI) Applied Research STEM Curriculum Unit Portfolio; Unit: Earth's Energy Budget** – Nicole Dulaney, Allegra LeGrande and Matthew Pearce – https://www.giss.nasa.gov/ – Accessed: September 2020

the warming of the Earth has been most pronounced.[338] This is shown in Figure 17. Since the 1980s, vegetative growth has been particularly pronounced in the high northern latitudes, greater than 50°N.[339]

1880-2019 temperature change *Figure 17*

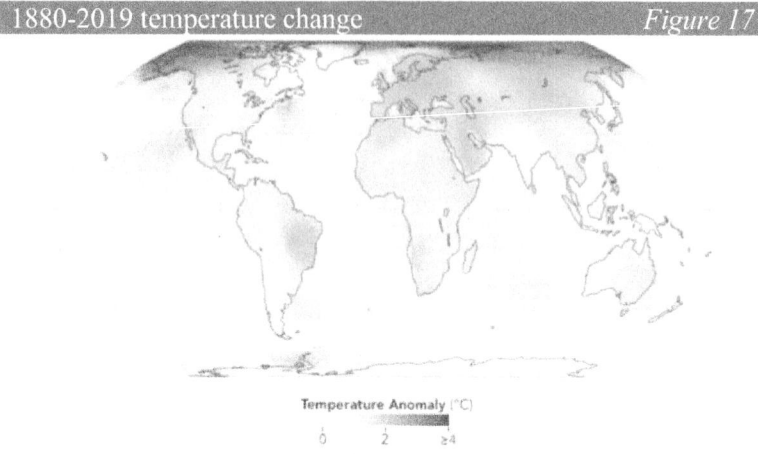

Image Credit: NASA Earth Observatory[340]

Science in Flux: Douglas Sheil, a globally recognized expert in the water dynamics of forests, summarizes, "We know enough to recognize that our picture of how the global climate system works is very different than how it was viewed even a few years ago … Expect surprises."[341]

Computer models have dominated climate science since 1990. However, in recent years, confident assertions based on computer derived outputs have been replaced by a recognition of the complexity of our Earth: In 2019, the most sophisticated computer

---

[338] **Interhemispheric Temperature Asymmetry over the Twentieth Century and in Future Projections** – Andrew R. Friedman, Yen-Ting Hwang, John C. H. Chiang and Dargan M. W. Frierson – Journal of Climate – 1 August 2013
[339] **Characteristics, drivers and feedbacks of global greening** – Xuhui Wang, Taejin Park, Chi Chen and Xu Lian – Nature Reviews Earth & Environment – December 2019
[340] **World of Change: Global Temperatures** – NASA Earth Observatory – https://earthobservatory.nasa.gov – Accessed February 2021
[341] **Forests, atmospheric water and an uncertain future: the new biology of the global water cycle** – Douglas Sheil – Forest Ecosystems – 20 March 2018

models that incorporate the effects of the Earth's changing vegetative landscapes into climate models are "are still inconclusive."[342]

Confounded by complexity: Trees affect the climate. The climate affects trees. Changes to either affect the other.[343]

Everything that affects trees also affects the climate. By extension, the regional scale burning of wildfires lit by hunter-gatherers to maintain grasslands as well as agricultural yields affect our climate – both affect the amount of the Earth that is treed.[344]

The science of the Earth becomes more complicated the more we learn.

Yet, taking inspiration from a founder of science itself, Isaac Newton, "Truth is ever to be found in simplicity, and not in the multiplicity and confusion of things." The core inescapable truths are follows: The Earth's green and forested areas are growing; carbon dioxide is fertilizing plant growth; changes to the Earth's vegetation are affecting the climate.

New Paradigm: By construction, the global warming paradigm that has existed since 1990 excludes the fertilization effect of carbon dioxide and the importance of human-caused landscape changes on the climate. Table 10 provides a measure of the degree to which that paradigm ignores the most profound changes occurring on our Earth, namely, its greening and forestation.

---

[342] **Characteristics, drivers and feedbacks of global greening** – Xuhui Wang, Taejin Park, Chi Chen and Xu Lian – Nature Reviews Earth & Environment – December 2019
[343] **Forests, atmospheric water and an uncertain future: the new biology of the global water cycle** – Douglas Sheil – Forest Ecosystems – 20 March 2018
[344] **Indigenous impacts on North American Great Plains fire regimes of the past millennium** – Christopher I. Roos, María Nieves Zedeño, Kacy L. Hollenback and Mary M. H. Erlick – Proceedings of the National Academy of Sciences of the United States of America – 7 August 2018

Changes: CO_2, greening, forests & temperature — *Table 10*

	Increase per Decade	Period
Atmospheric CO_2	5.1%	1975-2019
Forested area of Earth (net of losses)	2.0%	2000-2017
Green area of Earth (net of losses)	2.3%	1982-2016
Global surface temperatures	0.15-0.20°C (0.27-0.36°F)	1975-2019

Sources[345]

John Tyndall discovered greenhouse gases in 1859 and his discovery remains central to our understanding of the Earth's climate. However, the unexpected discoveries made in 2016, 2018 and 2019 relating to the rate of growth in the Earth's forested and green surface areas, as a result of the fertilization effect of carbon dioxide, are *by far* the most important discoveries ever made in relation to how carbon dioxide affects our Earth.

It is impossible to understate the profound debt science owes to NASA and its Terra satellite.

The global warming paradigm that emerges as a result of these changes will likely be unrecognizable from the paradigm that has existed for the last three decades.

The last word: At the time of writing, the most advanced research suggests that increased vegetation on Earth might have the effect of cooling the Earth, heating the Earth, or have an effect that depends on circumstances.[346]

---

[345] **Monthly Average Mauna Loa CO_2** – US National Oceanographic and Atmospheric Administration – https://www.esrl.noaa.gov – Accessed: March 2021;
Global land change from 1982 to 2016 – Xiao-Peng Song, Matthew C. Hansen, Stephen V. Stehman, Peter V. Potapov, Alexandra Tyukavina, Eric F. Vermote & John R. Townshend – Nature – 8 August 2018;
China and India lead in greening of the world through land-use management – Chi Chen, Taejin Park, Xuhui Wang, Shilong Piao, Baodong Xu, Rajiv K. Chaturvedi, Richard Fuchs, Victor Brovkin, Philippe Ciais, Rasmus Fensholt, Hans Tømmervik, Govindasamy Bala, Zaichun Zhu, Ramakrishna R. Nemani & Ranga B. Myneni – Nature Sustainability – 11 February 2019; and
World of Change: Global Temperatures – NASA Earth Observatory – https://earthobservatory.nasa.gov – Accessed February 2021

[346] **Climate impacts of U.S. forest loss span net warming to net cooling** – Christopher A. Williams, Huan Gu and Tong Jiao – Science Advances – 12 February 2021;
Biophysical impacts of Earth greening largely controlled by aerodynamic resistance – Chi Chen, Dan Li, Shilong Piao, Xuhui Wang, Maoyi Huang, Pierre Gentine,

In February 2021, a globally recognized professor of forestry indicated that planting trees for the purposes of cooling the planet could be counterproductive.[347] However, this assessment is based only on the heat absorbed by forests due to their dark color.

For the time being, the greenhouse-gas impacts of increased water vapor in the atmosphere due to the Earth's growing forests and green areas – the only thing that matters – have not been a subject of scientific research.

A coherent and comprehensive climate change paradigm that incorporates the most significant changes occurring on our Earth, namely, the unexpected growth in its forests and green areas, has yet to emerge. It is, however, clear that the facts have changed, even if that has not been communicated to the general public.

Ramakrishna R. Nemani and Ranga B. Myneni – Science Advances – 20 November 2020;
Climate mitigation from vegetation biophysical feedbacks during the past three decades – Zhenzhong Zeng, Shilong Piao, Laurent Z. X. Li, Liming Zhou, Philippe Ciais, Tao Wang, Yue Li, Xu Lian, Eric F. Wood, Pierre Friedlingstein, Jiafu Mao, Lyndon D. Estes, Ranga B. Myneni, Shushi Peng, Xiaoying Shi, Sonia I. Seneviratne & Yingping Wang – Nature Climate Change – 22 May 2017; and
Satellites reveal contrasting responses of regional climate to the widespread greening of Earth – Giovanni Forzieri, Ramdane Alkama, Diego G. Miralles, Alessandro Cescatti – Science – 16 June 2017
[347] **More trees do not always create a cooler planet, Clark University geographer finds** – Christopher A. Williams – Clark University – 12 February 2021

4. From Fire: Politics

"When the facts change, I change my mind. What do you do, sir?" was famously quipped by John Maynard Keynes, one of the great economists of the last century. In contrast, how new facts will affect the objective of extinguishing humanity's use of fire, namely, fire from burning oil, natural gas and coal, is not obvious. A lot will depend on our underlying motivations to extinguish fire. Let us look at two peripheral motivations to extinguish fire before looking at the core motivation:

i) Many regions of the world import their fuel, which creates a dependency. In 2018, 58% of the European Union's energy was imported. In that year, 30% of the European Union's oil imports came from Russia and 40% of its natural gas imports also came from that country.[348] Free trade amongst countries is widely acknowledged to maximize prosperity by encouraging people everywhere to do what they do best relative to others around the world. However, political considerations weigh heavily on the European Union's energy dependency: According to the European Council, "Since March 2014, the EU has progressively imposed restrictive measures against Russia. The measures were adopted in response to the illegal annexation of Crimea and the deliberate destabilization of Ukraine" by Russia.[349] The European Union is therefore in a quandary: It would like to be more assertive with Russia, but it is also dependent on Russian energy. The European Union is motivated to reduce its consumption of oil and natural gas to reduce its dependency on energy imports. This goal is aligned with the goal of reducing carbon dioxide emissions.

ii) From 2010 to 2020, over $2,600 billion was invested in alternatives to oil, gas and coal. The businesses that are in receipt of the equivalent amount of revenues over that period

---
[348] **From where do we import energy and how dependent are we?** – European Commission – https://ec.europa.eu/ – Accessed: September 2020
[349] **EU restrictive measures in response to the crisis in Ukraine** – European Council – https://www.consilium.europa.eu/ – Accessed: September 2020

are motivated by the goal of reducing emissions of carbon dioxide, but also by profits.[350] Businesses globally are both aligning their strategies with governmental regulations and influencing those regulations.

iii) The core motivation to reduce carbon dioxide emissions from fire is to limit the amount by which the Earth's surface temperature increases in order to avoid the extreme risks associated with global warming as elaborated by the Intergovernmental Panel on Climate Change.

Although the underlying motivations to extinguish the human use of fire are important, the social and economic means by which fire is to be extinguished may be even more important. Let us consider two opposite and extreme means of stopping the human use of fire before looking at how in reality that objective is being progressed:

i) Based on a free market approach, fire would be extinguished by alternatives that are better, cheaper and more abundant. Based on this approach, rather than being a burden on society, the replacement of fire would represent a tremendous opportunity to create prosperity, increase agricultural yields, reduce poverty and protect wildlife.

ii) At the other extreme, some people believe that stopping the human use of fire "is totally impossible to do simply by tinkering with market mechanisms … it will require a massive expansion of state ownership and comprehensive economic planning … To halt climate change, we need an ecological Leninism."[351] As a historical digression, under Lenin's communist government, the Russian Famine of 1921-1922 was "the worst, both as regards the numbers affected and as regards mortality from starvation and disease, which has occurred in Europe in modern times."[352]

---

[350] **Clean Energy Investment Is Set to Hit $2.6 Trillion This Decade** – Will Mathis – Bloomberg NEF – 5 September 2019
[351] **To Halt Climate Change, We Need an Ecological Leninism, an Interview with Andreas Malm** – Andreas Malm – Jacobin – 15 June 2020
[352] **Report on economic conditions in Russia: with special reference to the famine of 1921-1922 and the state of agriculture** – League of Nations, Secretariat – 1922 (via University of Warwick Library Collections)

iii) In practice, due to the extreme risks associated with global warming as elaborated by the Intergovernmental Panel on Climate Change and the slow pace of progress with which fire would be replaced with alternatives under free market conditions, people have accepted restrictions on personal freedoms and free markets. Meanwhile, governments have become more authoritative in suppressing the human use of fire, namely, fire from oil, natural gas and coal, while promoting alternatives to fire.

Having looked at various motivations for, and means of, extinguishing fire, we will now assess underlying psychological considerations related to the science and politics of stopping the human use of fire.

Humans do not objectively gather evidence to form unbiased conclusions. Rather, humans tend to gather evidence that confirms their prior beliefs and values while ignoring or rejecting contradictory information.[353] This is referred to as confirmation bias.

Due to confirmation bias, data is not necessarily relevant for the formation of beliefs relating to how carbon dioxide emissions affect our Earth for two key reasons:

i) People who are convinced of scientific conclusions dismiss new data that challenges those conclusions.

ii) People who prefer strong governmental controls of social and economic activity tend to gather scientific, ecological and social evidence that supports a need for strong governmental controls of carbon dioxide emissions.

Humans are not only affected by confirmation bias; our cognition is also influenced by anxiety and fear.[354] Anxiety and fear are emotional responses that affect human cognition in the following ways:

i) Anxiety and fear cause us to lose rational, well-judged control of our threat detection systems, making us hyper-

---

[353] **What Is Confirmation Bias?** – Shahram Heshmat – https://www.psychologytoday.com – 23 April 2015 – Accessed: September 2020
[354] **Neurobiological Correlates of Cognitions in Fear and Anxiety: A Cognitive-Neurobiological Information Processing Model** – Stefan G. Hofmann, Kristen K. Ellard and Greg J. Siegle – National Institute of Health – 1 February 2013

sensitive to anything that could potentially be perceived as threatening, whether or not it actually is.[355]

ii) Anxiety and fear disrupt rational decision making and lead to poor choices.[356]

iii) Anxiety and fear cause us to i) overreact to uncertainty and ii) exaggerate threat significance and likelihood.[357]

Brené Brown, a research professor at the University of Houston, distilled a breadth of social research into a succinct conclusion: "Anxiety is contagious."[358]

Anxiety and fear are also heightened by uniquely focusing on risks related to our changing Earth.

The Intergovernmental Panel on Climate Change produces reports such as the most recent Fifth Assessment Report on Climate Change. That report is particularly focused on i) forward-looking risks related to climate change based on computer-derived climate models and ii) risk assessments based on observed changes to our Earth.

Wildfire risks feature prominently in that report: The Intergovernmental Panel on Climate Change determined with "very high confidence" that global warming is already causing "extreme" wildfires that are causing harm to humans and ecosystems.[359]

The statements made by the United Nations' Intergovernmental Panel on Climate Change linking global warming to extreme

---

[355] **How anxiety warps your perception** – Bobby Azarian – In Depth, Psychology, BBC – 29 September 2016; and
The Principles of Psychology – William James – Henry Holt and Company – 1890
[356] **How Does Anxiety Short Circuit the Decision-Making Process?** – Christopher Bergland – https://www.psychologytoday.com – 17 March 2016 – Accessed: September 2020; and
Anxiety Evokes Hypofrontality and Disrupts Rule-Relevant Encoding by Dorsomedial Prefrontal Cortex Neurons – Junchol Park, Jesse Wood, Corina Bondi, Alberto Del Arco and Bita Moghaddam – Journal of Neuroscience – March 2016
[357] **5 Ways Anxious Feeling Changes the Way We Think, Irrational responses to uncertainty** – Shahram Heshmat – https://www.psychologytoday.com – 23 April 2015 – Accessed: September 2020
[358] **The psychology of mass panic and collective calm** – Kate Raynes-Goldie – medicalpress.com – 9 April 2020 – Accessed: September 2020
[359] **Climate Change 2014, Synthesis Report** – Editors: Rajendra K. Pachauri, Leo Meyer and Core Writing Team – Fifth Assessment Report of the Intergovernmental Panel on Climate Change – 2015

wildfires have been picked up and relayed by a multitude of organizations and businesses. For example, Ørsted, the number-one ranked developer of offshore wind projects globally, stated in its 2019 annual report, "Science has clearly demonstrated the need to limit global warming to 1.5° Celsius to avoid uncontrollable effects of climate change, including more floods, *wildfires* and droughts, decrease in biodiversity and many other severe consequences."[360] Without always applying independent critical analysis, the media too has extensively associated global warming with increased wildfire risks (Figure 18).

Media portrayal of rising climate-caused wildfire risks *Figure 18*

Sources: The Guardian and TIME Magazine[361]

Although ostensibly intended to convey scientific analysis, the cited report of the Intergovernmental Panel on Climate Change, because it focuses extensively on risks, produces anxiety and fear, which are emotional responses. The emotions of the authors, editors and reviewers of that report, rather than rational analysis, may have also been determinant in producing that report's conclusions.

---

[360] **Ørsted Annual Report 2019** – https://orsted.com
[361] **'Relentless' climate crisis intensified in 2020, says UN report** – Damian Carrington – The Guardian – 19 April 2021; and
TIME Magazine Cover – Edward Felsenthal, Editor-in-Chief and CEO of TIME – 26 April 2021

A strictly rational analysis of the best three sources of observed changes to wildfire dynamics globally provides a different conclusion and one that does not produce anxiety and fear:

i) Satellite data indicates that over the 18-year period ending in 2015 the amount of surface area burned globally by wildfires *decreased* by 24.3%.[362]

ii) Extreme wildfires are caused by extended cold and wet periods – not by warmth.[363]

iii) The longest-running publicly available data series globally relating to wildfire burn areas is provided in Figure 19. It indicates that wildfire burn areas have *decreased* over the last century.

USA wildfire burn areas (1927-2020; km² per year) *Figure 19*

Source: US National Interagency Fire Centre – https://www.nifc.gov

---

[362] **A human-driven decline in global burned area** – N. Andela, D. C. Morton, L. Giglio, Y. Chen, G. R. van der Werf, P. S. Kasibhatla, R. S. DeFries, G. J. Collatz, S. Hantson, S. Kloster, D. Bachelet, M. Forrest, G. Lasslop, F. Li, S. Mangeon, J. R. Melton, C. Yue, J. T. Randerson – Science – 30 June 2017
[363] **Multi-Millennial Fire History of the Giant Forest, Sequoia National Park, California, USA** – Thomas W. Swetnam, Christopher H. Baisan, Anthony C. Caprio, Peter M. Brown, Ramzi Touchan, R. Scott Anderson & Douglas J. Hallett – Fire Ecology – 1 December 2009;
Giant Sequoias Yield Longest Fire History From Tree Rings – Tony Caprio – National Park Service, US Department of the Interior – https://www.nps.gov – Accessed: June 2020; and
The international nature of the Miramichi Fire – Alan MacEachern – The Forestry Chronicle – May/June 2014

In contradiction to the conclusions of the Intergovernmental Panel on Climate Change, as the Earth's surface temperature has been *rising* wildfire activity has actually been *falling*.

Moreover, from the perspective of conservationists, the conclusions of the Intergovernmental Panel on Climate Change promote an outdated and unhelpful negative perception of wildfires. A better understanding of wildfires would reflect that:

i) Many ecosystems require wildfires and benefit from their occurrence.

ii) Prior to European settlement, half of the surface area of the Lower-48 States of the United States burned at intervals of at least one fire every 12 years (on average).[364] In contrast, using ten years of data up to and including 2019, the implied average frequency of wildfires across the Lower-48 States is one fire every 352 years.[365] From the perspective of maintaining natural ecosystems that have existed for millennia, the greatest threat relating to wildfires is their absence.

iii) Wildfire dynamics are overwhelmingly influenced by direct human activity such as prescribed burning, wildfire suppression, the transformation of landscapes and the planting of non-native trees such as eucalyptus trees. 84% of wildfires in the United States are started by humans.[366]

To understand the changes occurring to our dynamic Earth will require us to disentangle our emotions from rational analysis. That is not something humans do particularly well in a context where anxiety and fear have been created.

---

[364] **Presettlement Fire Frequency Regimes of the United States** – Cecil C. Frost – Pages 70-81 in Teresa L. Pruden and Leonard A. Brennan (eds.). Fire in Ecosystem Management: Shifting the Paradigm from Suppression to Prescription. Tall Timbers Fire Ecology Conference Proceedings, No. 20. Tall Timbers Research Station, Tallahassee, FL. – 1998

[365] **National Interagency Fire Center** – https://www.nifc.gov – Accessed: July 2020

[366] **Human-started wildfires expand the fire niche across the United States** – Jennifer K. Balch, Bethany A. Bradley, John T. Abatzoglou, R. Chelsea Nagy, Emily J. Fusco and Adam L. Mahood – Proceedings of the National Academy of Sciences of the United States of America – 14 March 2017

Positively, research suggests that calmness, like anxiety, spreads from one person to the next.[367] Moreover, based on the research of psychologists and mental health practitioners focusing on positive, constructive goals is a means of defeating anxiety.[368]

---

[367] **The psychology of mass panic and collective calm** – Kate Raynes-Goldie – medicalpress.com – 9 April 2020 – Accessed: September 2020
[368] **Alleviating Worry with Positive Thoughts** – Samuel Hunley –www.anxiety.org – Accessed: September 2020

Fire: Summary and Conclusion

Fire, Our Genetic Dependency

In contrast to widely held beliefs, humans did not invent the use of fire. Fire was used by now-extinct species long before our species existed.[369]

Humans have been using fire to cook food since we have existed as a species, starting some 300,000 years ago. Most importantly, advances in science indicate that the human species is genetically dependent on fire for its survival: Human morphology and in particular our energy-consuming brains and small digestive systems reflect that we require fire and the cooked food it provides.[370] Cooked food provides significantly more energy, net of digestion, than uncooked food.[371] Cooked food and human morphology allow us to allocate energy away from digesting towards thinking.[372]

Advances in science over the last two decades suggest that without fire humans would not exist.[373]

Fire and Language

Humans won the evolutionary jackpot twice: firstly, by becoming genetically adapted not just to our natural world, but to fire – an

---

[369] **Microstratigraphic evidence of in situ fire in the Acheulean strata of Wonderwerk Cave, Northern Cape province, South Africa** – Francesco Berna *et al.* – – Proceedings of the Natural Academy of Sciences of the United States of America (PNAS) – 15 May 2012

[370] **Catching Fire: How Cooking Made Us Human** – Richard Wrangham – Profile Books – September 2009;
Control of Fire in the Palaeolithic, Evaluating the Cooking Hypothesis – Richard Wrangham – Current Anthropology – 16 August 2017 and
Food for Thought: Was Cooking a Pivotal Step in Human Evolution? The dietary practice coincided with increases in brain size, evidence suggests – Alexandra Rosati – Scientific America – 26 February 2018

[371] **Cooking shapes the structure and function of the gut microbiome** – Rachel N. Carmody *et al.* – Nature Microbiology – 30 September 2019

[372] **Catching Fire: How Cooking Made Us Human** – Richard Wrangham – Profile Books – September 2009

[373] **Catching Fire: How Cooking Made Us Human** – Richard Wrangham – Profile Books – September 2009;
Control of Fire in the Palaeolithic, Evaluating the Cooking Hypothesis – Richard Wrangham – Current Anthropology – 16 August 2017 and
Food for Thought: Was Cooking a Pivotal Step in Human Evolution? The dietary practice coincided with increases in brain size, evidence suggests – Alexandra Rosati – Scientific America – 26 February 2018

invention – and, secondly, by the genetic acquisition of language. No other living species has either of these genetic attributes, and humans have both.

Our acquisition of fire and language explains why humans are so different from the other animals with whom we cohabitate our planet and why the course of human development is unlike anything else that has ever occurred on our Earth.

Fire and Farmland

Through the active lighting of wildfires our nomadic ancestors had an inordinate influence on many ecosystems globally.[374] In contrast, sedentary agricultural societies suppress wildfires.[375] The boundary between farmland and wilderness also represents the boundary between the wildfire regimes of sedentary and nomadic people.

The surprising 24.3% *decrease* in global wildfire burn areas over the 18-year period ending in 2015 is a reflection of the scale of the transformation of wilderness into farmland occurring on our Earth.[376]

From the time of the Natufians some 12,500 years ago, people have turned wilderness into farmland to address hunger and desperation. Today, hunger and desperation are concentrated in the Tropics;[377] as

---

[374] **Evolution of human-driven fire regimes in Africa** – Sally Archibald, A. Carla Staver and Simon A. Levin – Proceedings of the National Academy of Sciences of the United States of America – 17 January 2012;
Large Scale Anthropogenic Reduction of Forest Cover in Last Glacial Maximum Europe (Fires set by Ice Age hunters destroyed forests throughout Europe) – Jed O. Kaplan, Mirjam Pfeiffer, Jan C. A. Kolen, Basil A. S. Davis. – PLOS ONE – 1 December 2016; and
References on the American Indian Use of Fire in Ecosystems – Gerald W. Williams – United States Department of Agriculture, Forest Service – 18 May 2005 – https://www.nrcs.usda.gov – Accessed: February 2021
[375] **Researchers Detect a Global Drop in Fires** – Kate Ramsayer – NASA Earth Observatory – https://earthobservatory.nasa.gov – Accessed: June 2020
[376] **A human-driven decline in global burned area** – N. Andela, D. C. Morton, L. Giglio, Y. Chen, G. R. van der Werf, P. S. Kasibhatla, R. S. DeFries, G. J. Collatz, S. Hantson, S. Kloster, D. Bachelet, M. Forrest, G. Lasslop, F. Li, S. Mangeon, J. R. Melton, C. Yue, J. T. Randerson – Science – 30 June 2017
[377] **The world by latitudes: A global analysis of human population, development level and environment across the north–south axis over the past half century** – Matti Kumma and Olli Varis – Applied Geography – April 2011; and

a result, the expansion of farmland and the corresponding destruction of wildlife habitat are most pronounced in the Tropics.[378]

The African elephant population has fallen by a factor of 20 times since 1930 because African elephants have lost their habitat – their land.[379] In contrast to widely held perceptions, for wildlife global warming is irrelevant relative to the threat of habitat loss. Habitat loss is *by far* the greatest threat to wildlife globally.[380] Positively, when wildlands are recovered and given back to wildlife, wildlife populations recover. This is evidenced by the remarkable recovery in North American bison from near-extinction to a population of 350,000.[381]

Fire and Industry: Saviors of Wildlife

In contrast to widely held perceptions, from the time of the Industrial Revolution, fire, namely, fire from burning coal, oil and natural gas, has contributed significantly to providing wildlife with habitat because:

i) Fire from burning coal, oil and natural gas reduces demand for woodfuel, which benefits forests.

ii) Fire from burning coal, oil and natural gas provides tremendous amounts of energy relative to the minimal

---

State of the Tropics 2020 Report – Sandra Harding, Ann Penny, Shelley Templeman, Madeline McKenzie, Daniela Tello Toral and Erin Hunt – State of the Tropics – 2020
[378] **Global land change from 1982 to 2016** – Xiao-Peng Song, Matthew C. Hansen, Stephen V. Stehman, Peter V. Potapov, Alexandra Tyukavina, Eric F. Vermote & John R. Townshend – Nature – 8 August 2018; and
Forests, Trees, and Woodlands in Africa – Africa Region, World Bank – 11 October 2012
[379] **The Status of African elephants** – World Wildlife Fund Magazine – Winter 2018 – https://www.worldwildlife.org – Accessed: January 2021
[380] **Nature's Dangerous Decline 'Unprecedented' Species Extinction Rates 'Accelerating'** – United Nations, Intergovernmental Science-Policy on Biodiversity and Ecosystem Service – 6 May 2019;
"**Habitat Loss Poses the Greatest Threat to Species" Habitat Loss** – World Wildlife Fund – https://wwf.panda.org/ – Accessed: July 2020; and
Threats to Wildlife – The National Wildlife Federation – https://www.nwf.org/ – Accessed: July 2020
[381] **Meet the American Bison** – The Nature Conservancy – https://www.nature.org – Accessed 24 January 2021

amount of land required to source that energy. This frees land for wildlife.

iii) Fire from burning coal provides steel and concrete, which replaces wood use and reduces pressures on forests.

iv) Oil provides synthetic materials that replace agriculturally derived materials, such as cotton, which require land taken from wildlife.

v) Most importantly, fire from burning coal, oil and natural gas increases agricultural yields, which reduces land needs for agriculture. This increases the amount of land available for wildlife.

From Fire: Increasing Agricultural Yields

Increasing agricultural yields is the most time-proven means of increasing prosperity,[382] reducing hunger[383] and increasing the availability of land for wildlife.

According to the World Bank, "Agricultural development is one of the most powerful tools to end extreme poverty, boost shared prosperity and feed a projected 9.7 billion people by 2050."[384]

The International Monetary Fund's data shows a strong and direct relationship between agricultural yields and prosperity across countries.[385]

The human use of fire has been the driving force behind rising agricultural yields since bronze-reinforced ploughs were used during the Bronze Age. Figure 20 shows the evolution of wheat yields in the United Kingdom from the 1300s to the present.

---

[382] **Ending Extreme Poverty** – Interview of Ana Revenga by Amy Frykholm – 8 June 2016 – The World Bank (as first published by the Christian Century)
[383] **Ending Extreme Poverty** – Interview of Ana Revenga by Amy Frykholm – 8 June 2016 – The World Bank (as first published by the Christian Century)
[384] **Agriculture and Food** – The World Bank – https://www.worldbank.org – Accessed: March 2021
[385] **Crop Selection and International Differences in Aggregate Agricultural Productivity** – Jorge A. Alvarez and Claudia N. Berg – IMF Working Paper – August 2019

Evolution of wheat yields in the UK (1300-2019) *Figure 20*

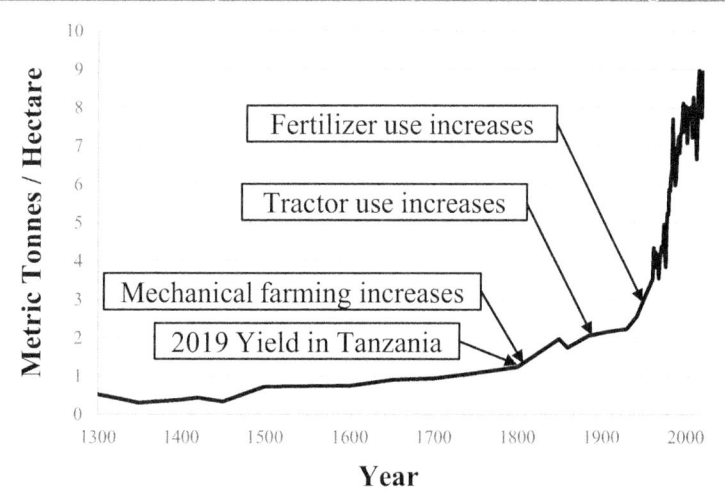

[Figure 20: Line chart with x-axis "Year" from 1300 to 2000 and y-axis "Metric Tonnes / Hectare" from 0 to 10. Annotations: "Fertilizer use increases", "Tractor use increases", "Mechanical farming increases", "2019 Yield in Tanzania".]

Sources[386]

Fire from burning coal fueled the Industrial Revolution and currently provides steel for farm equipment with low social burdens. Fire from burning oil provides motion to tractors and farm equipment with low social burdens. Natural gas and coal provide feedstock for fertilizers with low social burdens. The human use of fire has created the prosperity that provides the infrastructure, such as good roads, that supports rising agricultural yields.

Tanzanians cannot afford mechanized steel farm equipment, energy to provide motion to agricultural equipment or fertilizer. Figure 20 shows that Tanzania's wheat yield in 2019 was comparable to the United Kingdom's wheat yield in the early 1800s. Tanzania is 58 times poorer than the United States.[387] Tanzania's low agricultural

---

[386] **British Economic Growth, 1270-1870: an output-based approach** – Stephen Broadberry, Bruce Campbell, Alexander Klein and Mark Overton and Bas Van Leeuwen – British Economic Growth – December 2011;
Output and technical change in twentieth-century British agriculture – Paul Brassley – The Agricultural History Review – 2000;
Food and Agricultural Organization of the United Nations – FOAStat – http://www.fao.org – Accessed: March 2021; and
Tanzania Grain and Feed Annual 2019 – Benjamin Mtaki – US Department of Agriculture Foreign Agricultural Service – 9 April 2019
[387] **GDP per Capita, 2019** – World Bank – https://data.worldbank.org – Accessed: November 2020

yields are causing its treasured wildlife to be imperiled by the expansion of farmland.

Americans and other wealthy people might ask: Would you rather your country becomes more like Tanzania or would you rather Tanzania becomes more like your country? Increasing the social burdens of producing basic materials, energy and fertilizer will make them unaffordable for more people. That would have the effect of lowering global agricultural yields.

Well-intentioned efforts to suppress wildfires for the sake of saving Bambi were based on incomplete understandings and, in reality, had the effect of depriving mule deer of their food, which contributed to a decline in mule deer populations. Well-intentioned efforts to save wildlife by replacing fire with inferior alternatives would have the effect of decreasing agricultural yields. Given the scale and rate of growth in the human population, reducing agricultural yields would have the effect of displacing the Earth's remaining wilderness with farmland.

Positively, the human use of fire can be used to further increase global agricultural yields and thereby i) reduce extreme poverty ii) provide land for wildlife to flourish and iii) create prosperity.

In contrast to widely held perceptions, the most important social, economic and ecological impacts from replacing fire with alternatives to fire will relate to how that change affects agricultural yields.

Satellites: It's Good to See

Satellite data is providing us with a better understanding of our changing Earth. Rather than compounding the anxiety and fear created by the Intergovernmental Panel on Climate Change, the data is:

i) providing a basis for thoughtful reflection;
ii) dispelling the risks that have been propagated by that panel; and

iii) contradicting the core assumptions upon which the global warming paradigm has been established.[388]

New Global Warming Paradigm

The pace of change in climate science is difficult to exaggerate and will inevitably result in the emergence of a new global warming paradigm that reflects that carbon dioxide is fertilizing plant growth and that changes to the Earth's vegetative landscapes affect the climate.

Misjudged Threats

The assertion from the highest scientific authority on climate change with "very high confidence" that global warming is causing extreme wildfires that are already harming humans and ecosystems is unfounded and misleading.[389] Moreover, from the perspective of conservationists, the conclusions of the Intergovernmental Panel on Climate Change promote an outdated and unhelpful negative perception of wildfires.

Currently, 820 million people are being harmed, not by wildfires, but by malnourishment,[390] which affliction is concentrated in the Tropics. Harm caused by ongoing hunger includes the stunting of 144 million children under the age of five, implying their physical and mental development will be damaged permanently due to a lack of food.[391] That figure represents 21.3% of young children

---

[388] **China and India lead in greening of the world through land-use management** – Chi Chen, Taejin Park, Xuhui Wang, Shilong Piao, Baodong Xu, Rajiv K. Chaturvedi, Richard Fuchs, Victor Brovkin, Philippe Ciais, Rasmus Fensholt, Hans Tømmervik, Govindasamy Bala, Zaichun Zhu, Ramakrishna R. Nemani & Ranga B. Myneni – Nature Sustainability – 11 February 2019

[389] **Climate Change 2014, Synthesis Report** – Editors: Rajendra K. Pachauri, Leo Meyer and Core Writing Team – Fifth Assessment Report of the Intergovernmental Panel on Climate Change – 2015

[390] **The State of Food Security and Nutrition in the World 2019** – Food and Agriculture Organization of the United Nations, International Fund for Agricultural Development, UNICEF, World Food Programme and World Health Organization – 2019

[391] **World Health Statistics 2020** – World Health Organization: monitoring health for the SDGs, sustainable development goals - 2020

globally.[392] Today, 45% of child deaths globally are associated with malnutrition according to the World Health Organization.[393]

Three Decades of Failure

The first Scientific Assessment of Climate Change was published by the Intergovernmental Panel on Climate Change in 1990. Since that publication three decades ago, the goals of eradicating poverty and hunger and of protecting wildlife have not been prioritized relative to the goal of reducing atmospheric emissions of carbon dioxide. During that time:

i) There have been no visible or obvious reductions to global carbon dioxide emissions (see Figure 13),[394] despite the extraordinary social burdens incurred to achieve that goal.[395]

ii) The number of people suffering from hunger, which affliction is concentrated in the Tropics, has remained unchanged.[396]

iii) Wildlife habitat and, correspondingly, wildlife have been destroyed on a colossal scale, principally in the Tropics.[397]

iv) Many government-led policies that have been premised on the goal of reducing carbon dioxide emissions and protecting wildlife have directly contributed to increased carbon

---

[392] **Child Stunting** – World Health Organization – https://www.who.int – Accessed: March 2021

[393] **Malnutrition** – World Health Organization – https://www.who.int – Accessed: March 2021

[394] **Monthly Average Mauna Loa CO_2** – US National Oceanographic and Atmospheric Administration – https://www.esrl.noaa.gov – Accessed: August 2020

[395] **The Success of Wind and Solar is Powered by Strong Policy Support** – Laura Cozzi, Tim Gould and Paolo Frankl – The International Energy Agency – 1 June 2017; and
Clean Energy Investment Is Set to Hit $2.6 Trillion This Decade – Will Mathis – Bloomberg NEF – 5 September 2019

[396] **The State of Food Security and Nutrition in the World 2019** – Food and Agriculture Organization of the United Nations, International Fund for Agricultural Development, UNICEF, World Food Programme and World Health Organization – 2019

[397] **Global land change from 1982 to 2016** – Xiao-Peng Song, Matthew C. Hansen, Stephen V. Stehman, Peter V. Potapov, Alexandra Tyukavina, Eric F. Vermote & John R. Townshend – Nature – 8 August 2018; and
Forests, Trees, and Woodlands in Africa – Africa Region, World Bank – 11 October 2012

dioxide emissions and the large-scale destruction of wildlife.[398]

Defeating Anxiety with Positive Goals

"Nothing in life is to be feared, it is only to be understood. Now is the time to understand more, so that we may fear less." – Marie Curie

We are not by nature an anxiety ridden species. We are justified to be proud of our achievements inclusive of our mastery and use of fire. We have every reason to look to our future and that of our Earth with confidence.

Humanity's focus on the positive goals of alleviating hunger, creating prosperity and ensuring wildlife flourishes have been overwhelmed with negativity and anxiety caused by the extensive proliferation of threats, such as misleading statements related to growing wildfire risks.[399]

Achieving humanity's core underlying positive social, economic and ecological goals is, currently, incompatible with extinguishing the human use of fire.

"When it is obvious that the goals cannot be reached, don't adjust the goals, adjust the action steps." – Confucius

Fire is not Forever

In the coming decades, for many applications – not all – the replacement of fire by better alternatives is possible based on the rapid technological progress being made by alternatives to fire.

---

[398] **Globiom: the basis for biofuel policy post 2020** – Transport and Environment – April 2016;
Electric Mobility and Climate Protection: A Substantial Miscalculation – Ulrich Schmidt – Kiel Institute for the World Economy – June 2020; and
Around half of EU production of crop biodiesel is based on imports, not crops grown by EU farmers, new analysis – Transport and Environment – 16 October 2017
[399] **Climate Change 2014, Synthesis Report** – Editors: Rajendra K. Pachauri, Leo Meyer and Core Writing Team – Fifth Assessment Report of the Intergovernmental Panel on Climate Change – 2015

The development of superior alternatives to fire will i) increase prosperity, ii) increase agricultural yields and iii) reduce global poverty and hunger.

In the longer-term, if the pace of technological progress is maintained, the human use of fire will inevitably be replaced entirely by better alternatives.

That change will represent a much more significant transition than the transition between the Bronze Age and the Iron Age. We saw that between the Bronze Age and the Iron Age human civilization collapsed because the technology to make steel was not sufficiently advanced to compensate for the absence of bronze. It is worth considering that fire is much more fundamental to our survival, prosperity and development than bronze – or steel. Relative to fire, bronze and steel are mere by-products.

The replacement of fire with better alternatives will be comparable in significance to the Agricultural Revolution. The Agricultural Revolution forever changed how we source our food. The replacement of fire with better alternatives will forever change how we source our energy.

Conclusion

At the present time, there is no alternative to fire that is comparably useful, abundant or available with low social burdens – costs (see Table 6).

The *premature* replacement of fire with inferior substitutes that i) reduce prosperity ii) reduce agricultural yields and iii) increase global poverty and hunger would represent the greatest ecological and humanitarian catastrophe experienced on our Earth since the emergence of the human species some 300,000 years ago.

Positively, in time, the replacement of fire with better alternatives would create a new era of human development. That change would be comparable in significance to the changes that resulted from farming and the Agricultural Revolution.

Which road forward?

"That depends a good deal on where you want to get to."[400]

The goals of i) protecting wildlife, ii) eradicating hunger and iii) generating prosperity are complimentary and achievable.

From surviving the ice ages to landing on the moon, fire has been our foremost ally in the achievement of our most challenging goals since our species has existed.

To achieve positive social, economic and ecological goals requires us to assess the entirety of the impacts of our actions. Whether it is better for humans to use fire, or not, can be considered in that context. That will focus attention on how our actions will impact agricultural yields. Whether agricultural yields rise or fall will, in turn, determine whether wildlife thrives or perishes and whether the number of people suffering from hunger falls or rises.

For the first time in human existence, satellite data is able to provide comprehensive global information relating to how our actions are impacting the Earth. That data is almost always completely surprising and contrary to expectations, which highlights exactly why it is so valuable and why understanding it has potential to contribute significantly to the achievement of positive social, economic and ecological goals.

---

[400] **Alice's Adventures in Wonderland** – Lewis Carroll

I hope you have found *Fire* thought-provoking and worthwhile. If you have, I would be most grateful if you might share it (www.fire-the-story.com) on social media or leave a review on Amazon.

With thanks,

Brendan

Printed in Great Britain
by Amazon